SEDIMENT TRANSPORT IN IRRIGATION CANALS: A NEW APPROACH

UNESCO-IHE LECTURE NOTE SERIES

Sediment Transport in Irrigation Canals: A New Approach

HERMAN DEPEWEG
UNESCO-IHE, Delft, The Netherlands

KRISHNA P. PAUDEL
Consolidate Management Services, Nepal

NÉSTOR MÉNDEZ V
Universidad Centro Occidental "Lisandro Alvarado", Barquisimeto, Venezuela

CRC Press
Taylor & Francis Group
Boca Raton   London   New York   Leiden

CRC Press is an imprint of the
Taylor & Francis Group, an **informa** business

A BALKEMA BOOK

CRC Press/Balkema is an imprint of the Taylor & Francis Group, an informa business

Typeset by MPS Limited, Chennai, India
Printed and bound in The Netherlands by PrintSupport4U, Meppel

Although all care is taken to ensure integrity and the quality of this publication and the information herein, no responsibility is assumed by the publishers nor the author for any damage to the property or persons as a result of operation or use of this publication and/or the information contained herein.

Published by: CRC Press/Balkema
                P.O. Box 11320, 2301 EH Leiden, The Netherlands
                e-mail: Pub.NL@taylorandfrancis.com
                www.crcpress.com – www.taylorandfrancis.com

British Library Cataloguing in Publication Data
A catalogue record for this book is available from the British Library

Library of Congress Cataloging-in-Publication Data

Depeweg, Herman.
   Sediment transport in irrigation canals : a new approach / Herman Depeweg, UNESCO-IHE, Delft, The Netherlands [and two others]. – First edition.
      pages cm. – (UNESCO-IHE lecture note series)
   ISBN 978-1-138-02695-7 (paperback : alk. paper) – ISBN 978-1-315-74750-7 (ebook)
1. Sediment transport–Mathematical models.   2. Irrigation canals and flumes.   I. Title.
II. Series: UNESCO-IHE lecture note series.
   TC175.2.D48   2014
   551.3'53–dc23
                                                            2014020977

ISBN: 978-1-138-02695-7 (Pbk)
ISBN: 978-1-315-74750-7 (eBook)

Table of Contents

List of Figures

List of Tables

Preface

A fair, reliable and equitable supply of water to the end users is one of the prerequisites of a well-designed and managed irrigation system, that should be based on the knowledge of the various related sciences such as agriculture, hydraulics, hydrology, irrigation engineering, including canals and structures, soil science, as well as economic, social, management, operational and maintenance aspects. One of the lesser known and understood aspects of canal design and system operation concerns the specific characteristics and behaviour of sediment and its impact on the daily operation and maintenance activities of an irrigation system.

Professor H. Schoemaker taught me at the Delft University of Technology for the first time about the various comprehensive aspects of sediment in irrigation canals and presented the many international research activities in this specific irrigation field, mainly based on the type of sediment in rivers. At Delft Hydraulics Laboratory the research into the sediment problems in rivers, harbours, estuaries and rivers has reinforced for me the importance of the knowledge of sediment transport in general.

The many assignments abroad clearly showed the problems of sediment in various irrigation systems; the clogging of canals and structures often hampered the operation of the system and a just supply to the farmers. Several irrigation systems on the Indonesian islands, which I discussed with Professor H. Schoemaker and Professor H. Vlugter, the sediment problems in the Tihama Plain of Yemen, the Gezira Scheme in Sudan and several systems in South America all proved that sediment transport in non-wide canals was still a problem that needed to be solved in irrigation engineering.

UNESCO-IHE has given me all the support to deepen my knowledge of sediment transport in irrigation canals and the enthusiastic encouragement of Professor Bart Schultz has been a key support in my guidance of and assistance to MSc and PhD students in their research in this specific field of irrigation. Also the considerable and valuable advice of my colleague Eugene Dahmen and his analysis of sediment transport in irrigation canals and the CANDES and SYSDES models that were developed by him presented a fruitful basis in my attempts to deepen my knowledge. Especially the research activities of Nestor Mendez and his first steps to develop the computer model SETRIC formed a sound basis for other MSc research to deepen and to widen the obtained knowledge by their theoretical and field studies. About 10 years ago Krishna Paudel started his research into the same subject and his field measurements and theoretical improvements resulted in his PhD thesis and in an improved SETRIC. During the past few years Safraz Munir has researched more specifically the sediment behaviour in non-wide irrigation canals with downstream flow control.

All this newly gained knowledge and experience are laid down in these lecture notes with the aim of presenting our knowledge of non-cohesive sediment transport in non-wide irrigation canals. Further research on the behaviour of colloidal sediment will widen the scope of this specific field.

I am grateful for all the support that I have received in the past and at present to conceive these lecture notes. Only with the extraordinary support of my wife Loes and her enthusiasm to continue have I been able to finalize this second edition of the lecture notes.

CHAPTER 1

Introduction

The transport of sediment in irrigation canals influences to a great extent the sustainability of an irrigation and drainage network. Unintentional or unwanted erosion and/or deposition of sediment in canals will not only increase the maintenance costs, but also lead to an unfair and inadequate distribution of irrigation water to the end users. Proper knowledge of the behaviour and transport of sediment in these canals will help to plan efficient and reliable water delivery schedules to supply water at the required levels, to have a controlled deposition of sediments, to estimate and arrange the required maintenance activities, and to determine the type of desilting facilities and their efficiency, etc.

The study of sediment transport is mainly focused on the sediment and erosion processes in irrigation canal networks. In view of maintenance activities the head works should be designed in such a way that they prevent or limit the entrance of sediment into canals. In addition, the design of the canal network should be based upon the transport of all the sediment to the fields or to specific places in the canal system, where the deposited sediment can be removed at minimum cost. Sedimentation should be prevented in canals and near structures, as it will hamper and endanger a correct irrigation management, the main objectives of which are to deliver irrigation water in an adequate, reliable, fair and efficient way to all the farmers at the required water level, at the right time and at the proper rate. Inadequate management will result in low efficiency and unnecessary loss of the already scarce resource.

Irrigation canals are usually designed upon the assumption that the water flow is uniform and steady and that the canals are able to carry the water and sediments to the fields. The design supposes that an equilibrium situation exists where the sediments and water entering into the irrigation network will be transported to the fields without deposition or erosion. However, a perfectly uniform and steady flow is seldom found. In the operation of irrigation systems the flow is predominantly non-uniform, with varying discharges and very often with a constant water level at the regulation points where the water is supplied to the offtakes. The sediment transport capacity of the canals greatly depends on the flow conditions, which are variable. Although the water flow can be modelled with a high degree of accuracy, sediment transport behaviour is only understood to a limited extent. The predictability of sediment transport equations and models in view of the quantity of sediment that needs to be removed is still rather poor. Computations of the effects of the non-equilibrium flow conditions on sediment transport are required to determine whether deposition and/or entrainment will occur and to assess the amount and distribution of the sediment deposition and/or entrainment along the canals. Mathematical modelling of sediment transport offers the possibility of estimating the distribution of sediment deposition or entrainment rates for a particular flow and a specific situation

The main criterion for a canal design is the need to convey different amounts of water at a fixed level during the irrigation season in such a way that the irrigation requirements are met. Furthermore, the design must be compatible with the sediment load of a particular location in order to avoid silting and/or scouring of the canal. The water supply should meet the irrigation requirements and at the same time the supply should result in the least possible deposition in and/or scouring of the canals. The design process becomes more complicated when canals are unlined and pass through alluvial soils.

The problems of design and maintenance of stable channels in alluvial soils are fundamental to all irrigation schemes (Raudkivi, 1990). Stable or regime canals present ideal conditions for non-scouring and non-silting throughout the irrigation network after one or several seasons of intensive operation. The search for the main characteristics of stable channels started with the work of Kennedy between 1890 and 1894 (Kennedy, 1895). Subsequently, different theories have been developed and are being used around the world. All of them assume uniform and steady flow conditions and try to find those canal dimensions that are stable for a given discharge and sediment load.

In the past the irrigation canals used to be designed for protective irrigation, where the government or an irrigation authority ran and maintained the system. The available water was spread over as large an area as possible. The canals carried an almost constant discharge without significant control structures and to some extent the assumption of a uniform and steady flow was realised. In these situations a prediction of the behaviour of the sediment in the canal could be made with some reliability. The growing need for reducing the continually increasing governmental investments in irrigation demands a more economical system, so that the farmers can pay for the operational and maintenance costs with the return that they get from irrigated agriculture. In addition, farmers would like to have a more reliable and flexible water delivery. Flexibility in the delivery system demands more frequent regulation of the water flow, which will create unsteadiness in the flow. This unsteadiness will make all the assumptions made in the original design imprecise and reduces the accuracy of the already poor sediment transport predictors.

Generally three methods for the design of irrigation canals are used: namely Lacey's regime method and the tractive force method for large irrigation schemes, and the permissible velocity method for irrigation systems in hilly areas. Sediment characteristics and coefficients are either simply borrowed from literature or chosen based on the experience of the designer. Hence there exists a large difference in the design parameters from scheme to scheme and even from canal to canal within the same scheme. Without incentives or obligations for any verification, most designers do not investigate or evaluate the performance of their design in the field.

At the moment, the improvement of the performance of existing schemes is more pressing than the development of new irrigation systems, especially in view of the high investment costs required for the construction and operation of new systems. Appropriate management of sediment in canal networks is one of the major challenges of the improvement works, as a major part of the available maintenance budgets is spent annually on the removal of the sediment deposited. This type of scheme imposes extra conditions on the designer as the canal slope, bed width, structure control, and management practices already exist and in most cases they cannot be changed due to economic resources and/or social considerations. Hence, the selection of an appropriate design philosophy and its applicability for these particular conditions is very important.

The design of irrigation canals is not as simple as normally perceived. It is the final product of a merger of complex and undetermined parameters such as water flows, sediment load, structure control and operation, and management strategies. No design packages for irrigation canals are available that deal with all the parameters at the same time. To simplify the design process some parameters are either disregarded or assumed to be constant, which consequently will lead to a less adequate design. Many failures and problems are caused by a design approach that pays insufficient attention to the operational aspects (FAO, 2003). Considering the aforementioned parameters involved in the canal design and their importance in view of the sustainability of the system, numerical modelling may be one option that can simultaneously simulate all the variables. However, the selection of a model to represent the system and its validity in the proposed environment will have a major influence on the results.

Developments in the knowledge of sediment transport in open canals have mainly been derived from natural channels such as rivers. So far sediment transport theories, the development of bed forms, resistance factors, etc. have been developed under assumptions applicable to the particular conditions encountered in rivers. Even though certain similarities between rivers and irrigation canals exist, the sediment concepts for rivers are not entirely applicable to irrigation canals. Most of these irrigation canals are man-made and the irrigation environment presents a number of typical problems that are rarely encountered in rivers. The need to control water levels and discharges in the upstream and/or downstream direction, the necessity to find an optimal cross section and the considerable influence of the side banks on the velocity distribution normal to the flow direction create some of the main differences in both kinds of channels. Other differences are the presence of a large number of flow control structures, the occurrence of submerged gate flow and overflow structures, and the distinct flow characteristics of inverted siphons and their multiple flow paths. Table 1.1 shows some of the main differences between rivers and irrigation canals.

Irrigation canals are different from natural rivers in terms of hydraulics and sediment characteristics. The computational environment for the modelling of water flows in irrigation canals with sediment transport is much more demanding than for river flows due to the extreme variability and unsteadiness of the flow, the presence of numerous hydraulic structures, dynamic gate movements and pump operations, and the existing topographical complexity. The methods for designing stable canals are only useful in very specific flow conditions. For large changes in discharge

Table 1.1. Characteristics of water flow and sediment transport in rivers and irrigation canals.

Water flow and sediment transport

Characteristics	Rivers	Irrigation canals
– Main function	Conveyance of water and sediment	Diversion, conveyance and distribution of water for agriculture
– Discharge	Not controlled; increasing in downstream direction	Controlled by operation rules; decreasing in downstream direction
– Alignment	Rarely straight, bends, sinusoidal meanders and braids	Straight, wide bends
– Topology	Convergent	Divergent
– Flow control	(Almost) no control structures	Several flow control structures for water level and discharge
– Water profiles	Generally without water level control: nearly steady flow	Water level control: gradually varied flow and unsteady flow
– Velocity distribution	Nearly uniform velocity distribution in lateral direction	Distribution greatly affected by side walls and side slope
– Froude number	Wide range	Restricted by the operation of flow control structures (Fr < 0.4)
– Width (B)/depth (h)	$B/h > 15$ (wide canals)	$B/h < 7$–8
– Lining	Alluvial river bed	Man-made canals: lined or unlined
– Sediment concentration	Wide range	Controlled at head works
– Sediment size	Wide range of sediment size	Fine sediment
– Size distribution	Graded sediment	Nearly uniform distribution
– Sediment material	River bed	External sources
– Sediment transportation	Suspended and bed load	Mainly suspended load
– Bed forms	Mostly dunes	Mostly ripples and mega-ripples
– Roughness	Skin and form friction	Form friction

and sediment inputs they are inadequate to describe the sediment transport process. Under such conditions the process of sediment transport can best be described by numerical modelling.

Irrigation canal networks form a complicated hydraulic system as they have to handle the motion of water and sediment as well as the mutual interaction of both motions. The sediment transported by the flowing water causes changes in the bed and on the sides of the canals and these changes will also influence the water movement. Hence, sediment transport and water flow are interrelated. However, the time scales of the flow of water and the sediment transport are different in canals and therefore, the two processes will be discussed separately to specify their particular properties and characteristics.

The objective of these lecture notes is a description of the recently developed and tested sediment transport concepts in irrigation canals and, therefore, only those hydraulic aspects necessary for a better understanding of the sediment transport will be discussed in Chapter 2. This chapter will present a synopsis of the main hydraulic principles in open canals together with a short description of the dimensionless numbers used in the sediment concepts and a classification of flow types including uniform and non-uniform flow theories. Also the basics of the flow distribution above hydraulically rough and smooth boundary layers will be reviewed.

Before discussing the concepts of sediment transport the main properties of sediment will be given. Sediments are fragmented material formed by the physical and chemical disintegration of rocks and they can be divided into cohesive and non-cohesive sediments. Non-cohesive sediments do not have physical-chemical interaction and their size and weight are important in view of their behaviour. The total sediment transport in rivers and canals can be divided in suspended load and bed load (see Figure 1.1).

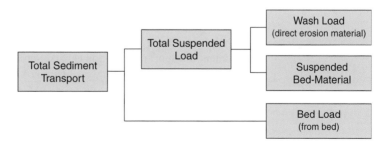

Figure 1.1. Classification of sediment transport.

Chapter 3 will present the main characteristics of sediments including density and porosity; particle size and size distribution; shape and fall velocity and the main dimensionless parameters used in the sediment transport theories.

Dahmen (1994) pointed out that an irrigation network should be designed and operated in such a way that the required flow passes at the design water level; no erosion of the canal bottom and banks occurs and no deposition of sediment in the canal takes place. The design of a canal requires a set of equations related to the water-sediment flow to provide the unknown variables of the bottom slope and cross section. The geometry of an irrigation canal that carries water and sediment will be the end result of a design process in which the flow of water and the transport of sediments interact. Chapter 4 outlines the main aspects of various design methods of irrigation canals, including the regime method, the tractive force and the permissible velocity approaches and the rational procedure.

Méndez (1998) developed the SETRIC model to improve the understanding of sediment transport in irrigation canals under changing flow conditions, sediment load and roughness conditions and for various operational and maintenance scenarios. SETRIC is a one-dimensional model, where the water flow in the canal has been schematised as quasi-steady and gradually varied. The one-dimensional flow equation is solved by the predictor-corrector method and Galappatti's (1983) depth integrated model has been used to predict the actual sediment concentration at any point under non-equilibrium conditions. Galappatti's model is based on the 2-D convection-diffusion equation. The mass balance equation for the total sediment transport is solved by Lax's modified method, assuming steady conditions for the sediment concentration. The model including the various calculation steps are presented in Chapter 6. For the prediction of the equilibrium concentration the following sediment predictors have been evaluated and the model uses the Brownlie, Engelund-Hansen and Ackers-White methods. These sediment transport concepts are extensively discussed and presented in Chapter 5 and Appendix A.

The SETRIC model can be applied to evaluate designs of an irrigation network and to analyse the alternatives, but it can also be used as a decision support tool in the operation and maintenance of a system and/or to determine the efficiency of sediment removal facilities in an irrigation system (Depeweg et al., 2002; 2003 and 2004). In addition the model can be helpful for the training of engineers to enhance their understanding of sediment transport in irrigation canals. Some examples of the application of SETRIC for the design and evaluation of irrigation canals are provided in Chapter 7.

In recent years SETRIC has been used to evaluate a large irrigation system in Nepal (Paudel, 2002) and to assess its suitability to predict non-equilibrium sediment transport (Ghimire, 2003). A special study analysed the applicability and versatility of the SETRIC model in an irrigation canal for different conditions of operation and sediment inputs (Sherpa, 2005). Also, a one-dimensional convection diffusion module has been developed to calculate the flow and sediment transport in both equilibrium and non-equilibrium conditions (Timilsina, 2005).

From the previous studies followed that the SETRIC model still has some limitations and the results obtained need careful and thorough verification. Paudel (2010) has used the model for a detailed analysis of field data and updated the design processes. The analysis and field measurements resulted in the necessary modifications and improvements, which has made the model more compatible with specific field conditions. The research of Mendez and Paudel has been mainly focussed on the water flow and sediment transport in irrigation networks with upstream flow controls. Munir (2011) has studied and investigated the role of sediment transport in the operation and maintenance of supply and demand based irrigation canals, which are operated under downstream control. A specific module for the sediment transport in downstream controlled canals has been included in SETRIC. The starting point of the model is the transport of non-cohesive sediment in the irrigation canal. A study has been initiated to investigate the behaviour of cohesive (colloidal) sediment in irrigation canals, which in the future might result in an extra module in SETRIC that handles the sediment transport of this type of sediment in irrigation networks.

A description and analysis of sediment transport concepts under the specific conditions of irrigation canals will contribute to an improved understanding of the behaviour of sediments in irrigation canals. It will also help to decide on improved water delivery plans in view of the operation and maintenance concepts. Finally it will help to evaluate design alternatives for minimal sedimentation and erosion, to maintain a fair, adequate and reliable water supply to the farmers, and to decide on the applicability of these concepts for the simulation of the sediment transport processes under the particular conditions of water flow and sediment inputs.

CHAPTER 2

Open Channel Flow

2.1 INTRODUCTION

Irrigation canals form a complicated hydraulic system as they have to handle the motion of water and sediment as well as the mutual interaction of both motions. Flowing water transports sediment, this sediment causes changes in the bed and on the sides of the canal, which also influences the water movement. Hence, sediment transport and water flow are inter-related and cannot be separated; they influence each other in an implicit manner. The time scales of the two processes are different (see Chapter 4) and therefore, the water flow and sediment transport can, in the first instance, be treated separately to specify their specific properties and characteristics.

However, to predict the topological changes in a canal the two phenomena also have to be considered in conjunction. The objective of these lecture notes is to present a description of the sediment transport concepts in irrigation canals and, therefore, only those hydraulic aspects that are necessary for a better understanding of the sediment transport will be discussed. This chapter will present a short overview of the main hydraulic principles in open channel flow, a description of the dimensionless numbers used in the sediment concepts and a classification of flow types. In addition, a summary of the uniform and non-uniform theories for steady and unsteady flows is included. For more information on the hydraulic theories, reference will be made to handbooks and lecture notes on hydraulics.

2.2 FLOW TYPES AND CHARACTERISTICS

Phenomena in open channels that convey water may vary considerably in magnitude and also sometimes in direction, both in terms of time and space. This chapter will introduce some main aspects of open channel hydraulics in order to be used as tools in the following chapters, which will deal with the movement of sediments and water together. In some cases the water flow varies over time and becomes unsteady. For some applications the variation may be considered to be so slow that a steady (or quasi steady) flow can be assumed. Considering the spatial distribution, any flow is essentially three-dimensional, meaning that the magnitude and direction of the flow vary from one point to another. Knowledge of this three-dimensional flow behaviour is still limited; but in many engineering applications it is often sufficient to know particular mean or average values.

The mean value can be presented for:

- a two-dimensional flow situation by averaging the value over the canal depth at a certain point; an example is the depth-averaged flow velocity in a vertical;
- a two-dimensional flow condition by averaging the value over the canal width (in a lateral direction); an example is the average water depth in a cross-section;
- a one-dimensional flow situation by averaging the value over the whole cross-section. The resulting values depend on the longitudinal coordinate (the *x*-values); an example is the average velocity in a cross-section.

It is important to remember that most of the two- and one-dimensional flow problems are simplified by averaging the flow characteristics; some information relating to the 'un-averaged' three-dimensional situation and the consequences of the process of averaging should be considered when the mean quantities are interpreted.

Open channel flow can be classified in many ways. A very common classification of open channel (gravity) flow is according to the change of flow depth with respect to time and space and is shown in Figure 2.1. Steady flow means that the main flow variables do not change and are steady with time; the variables at every point remain constant with time. An example is the flow in a channel where the water depth either does not change with time or can be assumed to be constant during the time interval

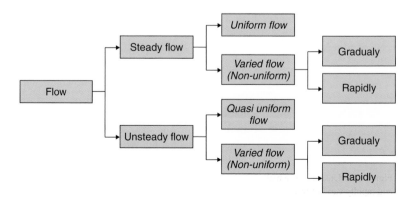

Figure 2.1. Classification of flow types.

under consideration. In unsteady flow the variables (magnitude and direction of the velocity, pressure, flow path, etc.) vary with time at the spatial points in the flow. Examples are surge waves in irrigation canals after a sudden opening or closing of a gate. In many open-channel problems it is sufficient to study the flow behaviour under steady conditions. However, when the change in flow condition with respect to time is of major concern, the flow should be treated as unsteady. In surges, for instance, the water level will change instantaneously as the waves pass by, and the time element becomes vitally important in the design of control structures.

In uniform flow the cross section (shape, side slope and area) through which water flows remains constant in the flow direction, the flow depth is the same in every section. In addition, the velocity does not change in magnitude and direction with distance. A uniform flow may be steady or unsteady, depending on whether or not the depth changes with time or during the time interval under consideration (see Figure 2.2). Steady uniform flow is the fundamental flow in open channel hydraulics. The establishment of unsteady, uniform flow would require that the water surface fluctuates from time to time while remaining parallel to the channel bottom. Obviously, this is a practically impossible condition.

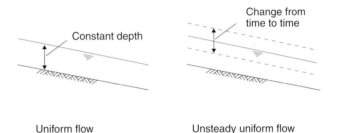

Figure 2.2. Steady and unsteady, uniform flow.

In varied flow, the cross section (shape, side slope and area) changes in the flow direction, the flow depth changes along the channel. Varied flow may be either steady or unsteady. Since unsteady uniform flow is rare, the term 'unsteady flow' is used exclusively for unsteady varied flow. Varied flow may be further classified as either rapidly or gradually varied. The flow varies rapidly if the depth changes abruptly over a relatively short distance; otherwise, it varies gradually. Figure 2.3 shows an example of unsteady, gradually and rapidly varied flow (G.V.F. and R.V.F.).

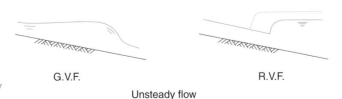

Figure 2.3. Unsteady, gradually and rapidly varied flow.

A rapidly varied flow is also known as a local phenomenon; examples are the hydraulic jump and the hydraulic drop. Spatially constant flow occurs when the average velocity is the same in all points; when the velocity changes along or across the flow, the flow is spatially variable. A clear example of a spatially varied flow is the flow through a gradual contraction or in a canal with a constant slope receiving inflow over the full length. Spatially varied flow shows some inflow and/or outflow along the reach under consideration; the continuity equation should be adapted to this situation. Examples are side channel spillways, main drainage canals and quaternary canals in irrigation systems. Figure 2.4 gives some typical locations in canals and rivers where gradually and rapidly varied flow might occur.

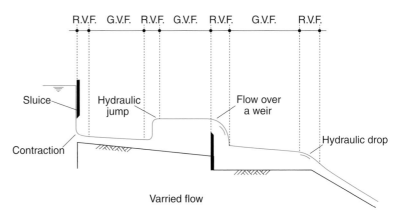

Figure 2.4. Examples of gradually and rapidly varied flow (steady conditions).

R.V.F. = Gradually varied flow

G.V.F. = Rapidly varied flow

Another flow classification is based on the dimensionless numbers that describe the relative influence of either the force due to viscosity or inertia in relation to the gravitational force. In hydraulics, the force due to viscosity is described as the resistance of a fluid to flow motion. The resistance acts against the motion of fluid when it passes fixed boundaries (e.g. the canal bottom or walls), but it also acts internally between slower and faster moving adjacent layers. Viscosity is the internal fluid friction that enables the acceleration of one layer relative to the other; it resists the motion of a layer but also makes it possible to accelerate a layer. Viscosity is the principal means by which energy is dissipated in fluid motion, typically as heat.

The difference in velocity between adjacent layers is known as a velocity gradient. To move one layer at a greater velocity than the adjacent layer, a force is necessary; resulting in a shear stress τ. Newton mentioned that for straight, parallel and uniform flow, the shear stress τ between layers is proportional to the velocity gradient, $\partial u/\partial y$, in the direction perpendicular to the layers.

$$\tau = \mu \frac{\partial u}{\partial y} \tag{2.1}$$

The constant μ is the dynamic viscosity. Fluids, such as water, that satisfy Newton's criterion are known as Newtonian fluids and show linearity between shear stress and velocity gradient. When the viscous forces are related to inertial forces, the ratio is characterized by the kinematic viscosity v.

$$v = \frac{\mu}{\rho} \tag{2.2}$$

The unit of dynamic viscosity is Pa·s (Pascal second) and is identical to $1 \, \text{N/s m}^2$. The unit of kinematic viscosity is m^2/s. The viscosity of water decreases with an increase in temperature, see Table 2.1.

Table 2.1. Kinematic viscosity of water as a function of the temperature T.

T (°C)	0	5	10	14	20	25	30
v (10^{-6} m^2/s)	1.79	1.52	1.31	1.14	1.01	0.90	0.80

Viscosity is the main factor resisting motion in laminar flow. However, when the velocity has increased to the point at which the flow becomes turbulent, pressure differences resulting from eddy currents rather than viscosity provide the major resistance to motion.

Significant dimensionless numbers in open channels are the Froude number and the Reynolds number. The Froude number gives the ratio between the inertial force and the force due to the gravity and is represented by:

$$\text{Fr}^2 = \frac{\rho L^2 v^2}{\rho g L^3} = \frac{v^2}{gL} \tag{2.3}$$

where:
$v =$ mean velocity (m/s)
$g =$ acceleration due to gravity (m/s^2)
$L =$ a characteristic length, e.g. water depth (m)
$\rho =$ density (kg/m^3)
$v =$ kinematic viscosity (m^2/s)

The Froude number differentiates between sub- and supercritical flow. When the Froude number is one, the flow is critical. For numbers larger than one the flow is supercritical.

The Reynolds number gives the ratio between the inertia force and the viscous force and is represented by:

$$Re = \frac{\rho v^2 L^2}{\mu v L} = \frac{vL}{v}$$

(2.4)

where:
v = mean velocity (m/s)
L = a characteristic length, e.g. water depth (m)
v = kinematic viscosity (m²/s)

For low Reynolds numbers the flow is laminar and for large numbers the flow is turbulent. An additional form of the Reynolds number is the number using the shear velocity:

$$Re_* = \frac{\rho u_* R}{\mu} = \frac{u_* R}{v}$$

(2.5)

where:
$u_* = \sqrt{gRS_o}$ = shear velocity (m/s)
g = acceleration due to gravity (m/s²)
R = hydraulic radius (m)
S_o = bottom slope (m/m)
v = kinematic viscosity (m²/s)

2.3 GEOMETRY

Canals are man-made channels and their cross-section can be prismatic (when A and S_o are constant) or non-prismatic. Moreover, the cross-section can be either regular (e.g. circular, triangular, rectangular or trapezoidal) or irregular. Figure 2.5 gives the main characteristics within a cross section and a longitudinal section of a canal.

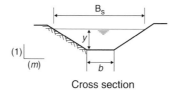

Cross section

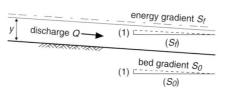

Figure 2.5. Main characteristics of a canal.

Longitudinal section

The cross-section is generally defined by the following geometrical measures.
- Water depth y: vertical distance from the bottom to the water surface
- Section depth d: normal (perpendicular) distance from the bottom to the water surface
- Surface or top width B_s: length of the channel width at the water surface
- Wetted area A: area normal to the flow direction
- Wetted perimeter P: length of the wetted line of intersection
- Hydraulic radius R: wetted area A/wetted perimeter P
- Hydraulic depth D: wetted area A/top width B_s
- Bottom width b: length of the bottom
- Side slope m: 1 vertical: m horizontal

Some other quantities that are related to a cross-section include:
- Discharge Q which is the total amount of water flowing through a cross-section during the unit of time t; the unit of discharge is m^3/s
- The average velocity is by definition: $v = Q/A$
- Total energy: $E = z + d \cos \Theta + \alpha v^2/2g$ ($y = d \cos \Theta$; Θ is the bottom slope in radian)
- Specific energy: $E_s = y + \alpha v^2/2g$ (for small slopes $\cos \Theta = 1$)
- Velocity head: $\alpha v^2/2g$ (α is the Coriolis coefficient)
- Froude number: $\mathrm{Fr}^2 = \alpha Q^2 B_s / g\, A^3$

2.4 BASIC HYDRAULIC PRINCIPLES

The motion of water can be described by considering the conservation of mass (continuity) and the conservation of momentum as expressed by Newton's second law. Sometimes the energy equation can also be helpful to describe the motion of water, especially when all the energy losses are known.

Continuity principle
The continuity equation is applicable for all flow types without any restriction. By using an x-, y-, z-coordinate system that is fixed in space and the velocity components u, v and w, respectively, the continuity equation is obtained by considering a small fluid element and the net rate of mass entering that element. The continuity equation for unsteady, compressible, real (with friction) or ideal (no friction) fluids reads:

$$\frac{\delta \rho u}{\delta x} + \frac{\delta \rho v}{\delta y} + \frac{\delta \rho w}{\delta z} + \frac{\delta \rho}{\delta t} = 0 \tag{2.6}$$

From this general, continuity equation follows the equation for three-dimensional, steady and incompressible flow (ρ = constant):

$$\frac{\delta u}{\delta x} + \frac{\delta v}{\delta y} + \frac{\delta w}{\delta z} = 0 \qquad (2.7)$$

In the most common case, namely a two-dimensional flow, the equation becomes:

$$\frac{\delta u}{\delta x} + \frac{\delta v}{\delta y} = 0 \qquad (2.8)$$

When $\delta v / \delta y = 0$, the flow is one-dimensional and the equation reads:

$$\frac{\delta u}{\delta x} = 0 \qquad (2.9)$$

This equation results in the fact that the velocity u is constant and it represents uniform flow; the direction and magnitude of the velocity in all points are the same. For a steady flow, the discharge Q is constant along the canal and the continuity principle reads:

$$Q = \int^{A} v \, dA = \bar{v} A = \text{constant}$$
$$Q = \bar{v}_1 A_1 = \bar{v}_2 A_2 \qquad (2.10)$$

where:
A = area of the cross section (m^2)
$\bar{v}$ = mean velocity perpendicular to the cross-section (m/s)

When the mean velocity in an open channel is constant ($v_1 = v_2$) then the area A of the cross sections is the same; the channel has a prismatic cross section with $A_1 = A_2$. When the flow is uniform, the water depth will be the same in all sections.

Conservation of momentum
The second law of Newton states that the sum of all external forces P equals the rate of change of momentum. The change of momentum per unit of time is equal to the resultant of all external forces (hydrostatic, friction, weight) acting on the body of flowing water:

$$\sum \vec{P} = \lim_{\Delta t \to 0} \frac{\Delta (m\vec{v})}{\Delta t} \qquad (2.11)$$

$$\sum \vec{P} = \frac{d(m\vec{v})_{\text{out}} - d(m\vec{v})_{\text{in}}}{dt} + \frac{(m_1 \vec{v}_1)_{t+\Delta t} - (m_1 \vec{v}_1)_t}{dt} \qquad (2.12)$$

The resultant of the external forces is equal to the rate of change of momentum of that body. *P* and *v* represent vectors; hence the change in momentum has the same direction as the resultant. Equation 2.12 states that the forces acting on a fluid mass are equal to the rate of change of the momentum of that mass. The first term on the right side represents the net flow rate of momentum going out of the control volume and the second term represents the rate of accumulation of momentum within the control volume during the time interval Δt.

The direction of ΣP is the same as that of Δv; ΣP represents the vectorial summation of all forces on the mass, including the gravity forces, shear forces, and pressure forces including those exerted by fluid surrounding the mass as well as the pressure forces exerted by boundaries in contact with the mass. The equation always applies.

In the case of steady flow, the last term in equation 2.12 is equal to zero, the force is equal to the net momentum outflow across the control surface and the equation becomes:

$$\sum \vec{P} = \frac{d(m\vec{v})_{\text{out}} - d(m\vec{v})_{\text{in}}}{dt} = \frac{d(m\vec{v})_{\text{out}}}{dt} - \frac{d(m\vec{v})_{\text{in}}}{dt} \tag{2.13}$$

$$\sum \vec{P} = \rho Q(\beta_2 \vec{v}_2 - \beta_1 \vec{v}_1) \tag{2.14}$$

where:
$\sum P =$ sum of all external forces acting on the control body (N)
$Q =$ discharge (m^3/s)
$A =$ area of the cross section (m^2)
$\vec{v} =$ mean velocity perpendicular to the cross-section (m/s)
$m =$ mass of water passing the cross section (kg)
$m = \rho v A$
$\beta =$ Boussinesq coefficient (see Section 2.5)

In this equation, the factor β is the coefficient of Boussinesq, which incorporates the effect of the velocity distribution on the average velocity. When there are no external friction forces, only the conditions at the end sections of the control volume govern the impulse-momentum principle.

Energy principle
The energy equation holds true as long as proper allowance is made for the energy losses. The total energy of a fluid particle per unit of volume is the sum of three types of energy (kinetic, potential and pressure). For an open channel with steady flow, and with straight and parallel streamlines,

the sum of the potential and pressure energy in the z-direction is:

$\rho g z + p = \text{constant}$. The kinetic energy of a particle is equal to $\dfrac{1}{2}\rho v^2$

The total energy for an entire cross section is equal to the sum of the energy of all the fluid particles. To find the total kinetic energy in this particular cross section, the velocity is expressed by the mean velocity: $v_{\text{mean}} = Q/A$. However, in open channels the velocity is not uniformly distributed over the depth and the width due to the presence of the free water surface and the friction along the boundary (bottom and sides).

For the total energy head this fact is taken into account by multiplying the velocity head ($v^2/2g$) by the coefficient α, the Coriolis coefficient (see Section 2.5). The true average kinetic energy across the cross-section per unit of volume is: $(\frac{1}{2}\rho v^2)_{\text{average}} = \frac{1}{2}\alpha\rho\bar{v}^2$.

The total energy E passing through a cross section is:

$$E = \frac{1}{2}\alpha\rho\bar{v}^2 + \rho g z + p \qquad (2.15)$$

The total energy E divided by the weight results in the energy head, being the total energy per unit weight.

$$E_{\text{tot}} = \frac{\alpha\bar{v}^2}{2g} + \frac{p}{\rho g} + z \qquad (2.16)$$

If you assume two cross sections perpendicular to straight and parallel streamlines, and that the energy loss is negligible (the energy principle of Bernoulli), then:

$$E_{\text{tot}} = \frac{\alpha_1 v_1^2}{2g} + \frac{p_1}{\rho g} + z_1 = \frac{\alpha_2 v_2^2}{2g} + \frac{p_2}{\rho g} + z_2 \qquad (2.17)$$

where:
z = height of the channel bottom in m above datum
$p/\rho g$ = pressure head in a cross-section in m
$\alpha v^2/2g$ = velocity head in a cross-section (kinetic energy) in m
E_{tot} = total energy head in a cross-section in m above a datum

In a *real fluid* there are always friction and/or local losses ($\Delta E = h_f$)

$$\frac{\alpha_1 \bar{v}_1^2}{2g} + \frac{p_1}{\rho g} + z_1 = \frac{\alpha_2 \bar{v}_2^2}{2g} + \frac{p_2}{\rho g} + z_2 + h_f \qquad (2.18)$$

Critical flow

For open channel flow the *specific energy* (Bakhmeteff, 1912) is the energy per unit weight with respect to the channel bottom. The piezometric head in a point of the cross section is $(p/\rho g) + z$, which is equal to the water depth y in that particular cross section: $(p/\rho g) + z = y$. Therefore, the specific energy E_s is the sum of the water depth and the velocity head (assuming that the streamlines are straight and parallel):

$$E_s = y + \alpha \bar{v}^2 / 2g \qquad (2.19)$$

$$E_s = y + \alpha Q^2 / 2gA^2 \qquad (2.20)$$

where:

y = water depth (m)
Q = discharge (m³/s)
A = area of the cross section (m²)
$\bar{v}$ = mean velocity in a cross-section (m/s)
E_s = specific energy with respect to the bottom (m)

For a given cross-section and a constant discharge Q, the specific energy E_s is a function of the water-depth y only. Plotting this water-depth y against the specific energy E_s gives a specific energy curve for the discharge Q (see Figure 2.6).

Specific energy E_s function of water depth y

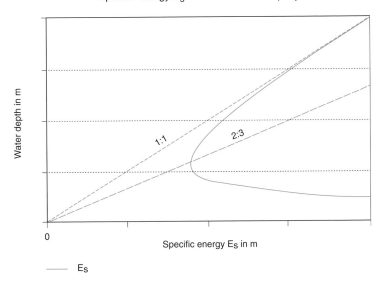

Figure 2.6. Specific energy as function of the water depth y.

From this figure with $E_s = y + Q^2 / 2gA^2$ follows that for a given discharge Q and a specific energy E_s, which should be larger than a certain threshold, there are *two alternate water depths*. For the threshold value, the specific energy is the minimal energy needed to convey that discharge Q through that cross section. In the case of the minimum energy the two alternate depths coincide for the given discharge Q and the water depth becomes the 'critical depth' (y_c). The Froude number for that discharge and water depth becomes 1 ($Fr^2 = 1$). In other words, for this critical depth the specific energy has a minimum value for the given discharge Q and cross section A.

When the flow depth is greater than the critical depth, the flow is subcritical; if it is less the flow is supercritical (see Table 2.2). For any discharge Q there are two possible flow regimes, but in reality only one will occur. These two flow regimes are either a slow and deep subcritical flow or a fast and shallow supercritical flow.

Table 2.2. Flow type as a function of the actual water depth.

Actual depth y in relation to y_c	Froude number	Flow type
$y > y_c$	$Fr^2 < 1$	Subcritical
$y = y_c$	$Fr^2 = 1$	Critical
$y < y_c$	$Fr^2 > 1$	Supercritical

For critical flow conditions:

- and for a given discharge Q the specific energy E_s has a minimum value;
- and for a given specific energy E_s the discharge Q has a maximum value;
- the velocity head is half the hydraulic depth D;
- the Froude number Fr^2 is one (unity).

2.5 VELOCITY DISTRIBUTION

In the next chapters it will be shown that the average velocity $v = Q/A$, and the deviation from this average velocity, are very important aspects in hydraulics and sediment transport. Water flowing in a cross section with a rigid or movable boundary (bottom and wall) will show a specific velocity distribution across that section. The velocity will be zero at the fixed boundary and next the velocity increases rapidly towards the middle and upper part of the channel. The velocity in a specific point is a function of the x, y and z coordinates (Figure 2.7).

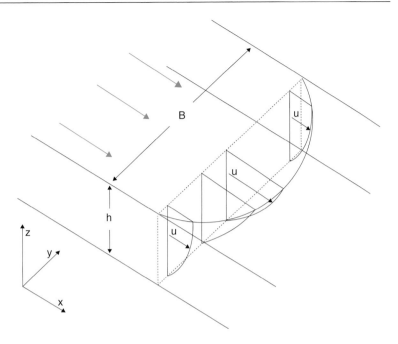

Figure 2.7. Distribution of the velocity in a rectangular channel.

The velocity profile in the vertical direction for a rectangular cross section can be approximated by the logarithmic equation as given by van Rijn (1984):

$$v_z = \frac{u_*}{\kappa} \ln\left(\frac{z}{z_0}\right) \tag{2.21}$$

where:
u_* = shear velocity (m/s)
v_z = the velocity at a point z (m/s)
κ = constant of von Karman (from measurements $\kappa = 0.4$)
z_0 = level of the zero velocity (m)
z = the level of a point above a datum (m)

By definition the shear velocity $u_* = \sqrt{\tau/\rho}$ and it can be expressed as:

$$u_* = \sqrt{\frac{\tau}{\rho}} = \sqrt{gRS_o} \tag{2.22}$$

By using $\kappa = 0.4$ the relationship for v_z becomes:

$$v_z = 2.5 u_* \ln\left(\frac{z}{z_0}\right) \tag{2.23}$$

The velocity v_z is equal to the average velocity $\bar{v}$ for $z = 0.4y$.

$$\bar{v} = 2.5u_* \ln\left(\frac{0.4y}{z_0}\right) = 5.75u_* \log\left(\frac{0.4y}{z_0}\right) \tag{2.24}$$

Results from measurements show that the logarithmic velocity profile gives a good approximation for the full depth of the flow due to a simultaneous decrease both in shear stress and mixing length l with z. Values of z_0 were determined from experiments with smooth and rough boundaries.

For smooth boundaries, a viscous sub-layer exists in which viscous effects predominate. The approximate thickness of this layer is $\delta = 10v/u_*$ and $z_0 = 0.01\delta = 0.1v/u_*$. For rough boundaries with uniform roughness Nikuradse has determined $z_0 = 0.03k_s$, in which k_s is the size of the sand particles used as roughness in the experiments. This k_s is also used as a standard roughness for other types of roughness (see Figure 2.8).

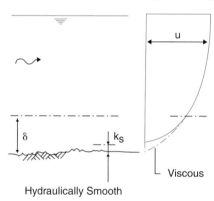

Hydraulically Smooth

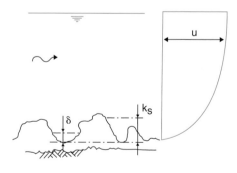

Hydraulically Rough

Figure 2.8. Hydraulically rough and hydraulically smooth boundary layers.

A boundary is:
- hydraulically rough for $k_s > 6\delta$
- hydraulically smooth $k_s < 0.1\delta$
- for intermediate values of k_s, the boundary is in a transition region.

White and Colebrook established an accurate good interpolation by taking $z_0 = 0.03k_s + 0.01\delta$. For this transition region the equation reads:

$$\bar{v} = 5.75u_* \log\left(\frac{12y}{k_s + 0.03\delta}\right) \tag{2.25}$$

The last equation can be transformed to the uniform flow equation of the de Chézy (see Section 2.6).

In the horizontal direction, the velocity distribution is influenced by the B/y ratio. For a non-wide canal ($B/y < 5$), the velocity distribution is three-dimensional. When the B/y ratio is larger than 5, the velocity distribution becomes almost two-dimensional with the exception of a small region near the vertical sidewalls.

Corrections in view of the average velocity

The velocity distribution in the vertical and width direction of a canal incorporates the fact that the velocity is not the same in all the points of the cross section. For that reason, the velocity in a point (v_A) is not equal to the average velocity $\bar{v} = Q/A$, but it can be written as $v_A = \bar{v} \pm \Delta v$. The discharge Q is by definition equal to the average velocity $\bar{v}$ times the total area A, but it is also the sum of the volumes of water passing the small areas ΔA with a velocity v_A.

$$Q = \sum v_A \Delta A = \sum (\bar{v} \pm \Delta v)\Delta A = \sum \bar{v}\Delta A \pm \sum \Delta v \Delta A$$

By definition: $Q = \sum (\bar{v}\Delta A)$ and hence: $\pm \sum (\Delta v \Delta A) = 0$

The total energy in a cross section comprises the potential energy, the pressure energy and the kinetic energy; the latter is expressed by the velocity head $v^2/2g$. The velocity head for the average velocity differs from the summation of the velocity head for each point and, therefore, it should be multiplied by the coefficient of Coriolis (α) to obtain the total head.

The mass of water flowing with a velocity v_a through a small area ΔA is $\rho v_a \Delta A$. The kinetic energy passing that area per unit of time is the product of the mass and the velocity squared: $\frac{1}{2}\rho v_a^3 \Delta A$ and the total kinetic energy for the whole cross section is: $\frac{1}{2}\sum \rho v_A^3 \Delta A$.

The total kinetic energy for the whole cross section can be expressed as $\alpha \rho A(\bar{v}^3/2g)$.

Equating this quantity with $\frac{1}{2}\sum \rho v_A^3 \Delta A$ results in

$$\alpha = \frac{\sum (v_A^3 * \Delta A)}{(\bar{v}^3 * A)} = \frac{\sum (\bar{v} \pm \Delta v)^3 \Delta A}{(\bar{v}^3 * A)}$$

$$\alpha = \frac{\sum (\bar{v}^3 \Delta A \pm 3\bar{v}^2 \Delta v \Delta A + 3\bar{v}\Delta v^2 \Delta A \pm \Delta v^3 * \Delta A)}{\bar{v}^3 * A}$$

As shown before $\sum \Delta v * \Delta A = 0$; also Δv^2 is always larger than zero and $\Delta v^3 * \Delta A$ can be ignored as it is very small. Hence, the coefficient of Coriolis can be presented as:

$$\alpha = 1 + 3 \frac{\sum \Delta v^2 * \Delta A}{\bar{v}^2 * A} \tag{2.26}$$

Another example of the effect of the velocity distribution on a hydraulic equation is the application of the second law of Newton in some hydraulic problems. The law states that the sum of all external forces is equivalent to the rate of change of momentum:

$$\sum \vec{P} = \Delta m \vec{v} = (m\vec{v})_{\text{out}} - (m\vec{v})_{\text{in}} \tag{2.27}$$

The mass of water flowing with a velocity v_a through an area ΔA is $\rho v_a \Delta A$. The momentum passing that area per unit of time is the product of the mass and the velocity: $\rho v_a^2 \Delta A$. The momentum for the whole cross section with area A is: $\sum \rho v_A^2 \Delta A$. The total momentum for the whole cross section can also be expressed by the average velocity $\bar{v}$ as: $\beta \rho \bar{v}^2 A$.

Equating this quantity with $\sum \rho v_A^2 \Delta A$ results in

$$\beta = \frac{\sum v_A^2 * \Delta A}{(\bar{v}^2 * A)}$$

$$\beta = \frac{\sum (\bar{v} \pm \Delta v)^2 \Delta A}{\bar{v}^2 * A} = \frac{\sum (\bar{v}^2 \pm 2\bar{v} * \Delta v + \Delta v^2) \Delta A}{\bar{v}^2 * A}$$

where $\sum \Delta v * \Delta A = 0$

$$\beta = 1 + \frac{\sum \Delta v^2 \Delta A}{\bar{v}^2 A} \tag{2.28}$$

where β = coefficient of Boussinesq (α is always larger than β).

2.6 UNIFORM FLOW

Uniform flow in open channels is characterized by:
- the depth, cross section, velocity and discharge are constant in every section;
- the lines that represent the energy, water surface and channel bottom are parallel; the slopes are $S_o = S_f = S_w$.

It is assumed that uniform flow in open channels is steady and turbulent (meaning that Re $\gg$ 600). Flow in open channels encounters friction and for uniform flow ($v = Q/A = $ constant) the component of gravity in the flow direction balances the friction forces. These two assumptions form the basis for the equations for uniform flow, including the de Chézy formula.

When water flows in a channel, a force is developed that acts on the boundary in the flow direction and is called the tractive force. In uniform flow, the tractive force is equal to the component of the gravity force acting on the control body parallel to the bottom (Simons & Sentruk, 1992).

The tractive force follows from the product of the shear stress and the contact area. Assume that the average shear stress on the perimeter P is τ. Next, the balance between the component of the gravity force and the frictional resistance (tractive force) is used to derive that average shear stress (see Figure 2.9). The shear stress τ follows from the weight component ($=\rho g A L S_o$) and the frictional resistance ($=\tau P L$):

$$\tau = \rho g R S_o \tag{2.29}$$

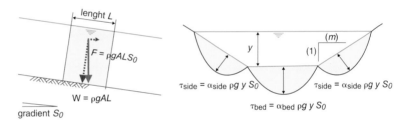

Figure 2.9. Tractive force and the distribution of the shear stress in a trapezoidal channel.

Tractive force on reach   Distribution of shear stress

By definition the shear velocity $u_* = \sqrt{\tau/\rho}$ and can be expressed as:

$$u_* = \sqrt{\frac{\tau}{\rho}} = \sqrt{gRS_o} \tag{2.30}$$

For fully turbulent flow Prandtl showed that the shear stress τ is a function of v^2 and the relationship between v and τ can be written as $\tau = Kv^2 = \rho g R S_o$. From this relation follows:

$$v^2 = \frac{\rho g}{K}RS_o = C^2 RS_o \tag{2.31}$$

This relation is normally presented as the de Chézy equation:

$$v = C\sqrt{RS_o} \tag{2.32}$$

Investigations in laboratories and measurements in the field have resulted in the following expression, proposed by White and Colebrook, for the coefficient C:

$$C = 18 \log \frac{12R}{k + \delta/3.5} \tag{2.33}$$

where:

$k =$ length characterizing the roughness (Nikuradse)
$\delta =$ thickness of the laminar sub-layer $= 11.6\nu/v_*$
$v_* = \sqrt{gRS_\mathrm{o}}$
$\nu = 10^{-6}\,\mathrm{m^2/s}$ (for $T = 20°C$)

Values of the de Chézy coefficient as function of the hydraulic radius R, the wall roughness k and the thickness of the laminar sub-layer δ are presented in Figure 2.10.

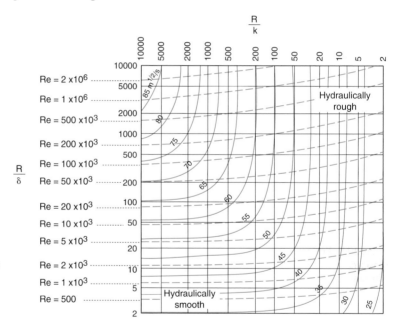

Figure 2.10. Values of the de Chézy coefficient C as function of the hydraulic radius R, the wall roughness k and the thickness of the laminar sub-layer δ.

Another, frequently used, equation for uniform flow is the Manning equation with n for the roughness:

$$v = \frac{1}{n} R^{2/3} S_\mathrm{o}^{1/2} \tag{2.34}$$

In some countries, the Strickler equation is preferred; this equation uses k_s for the roughness.

$$v = k_\mathrm{s} R^{2/3} S_\mathrm{o}^{1/2} \tag{2.35}$$

The relationship between C of de Chézy and Manning's n can be found by taking the same average velocity v.

$$C = \frac{R^{1/6}}{n} \tag{2.36}$$

2.7 NON-UNIFORM STEADY FLOW

A gradually varied flow is a steady flow, whose depth varies gradually along the channel:
- Hydraulic flow characteristics remain constant in time,
- Streamlines are practically parallel; hydrostatic pressure prevails,
- Bed friction is assumed to be equal to the friction in uniform flow (e.g. as used in the equations of Manning and de Chézy).

Figure 2.11 gives the energy considerations for unsteady flow in open channels, which also can be used for gradually varied flow (steady flow with $dv/dt = 0$).

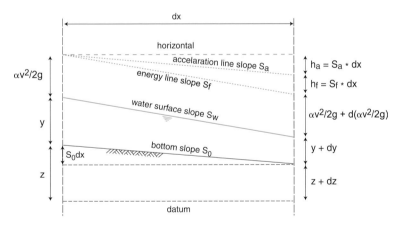

Figure 2.11. Energy considerations in a channel with unsteady flow.

The assumptions made in deriving the gradual varied flow equations are (Cunge et al., 1980):
- the flow is one dimensional;
- the flow velocity is uniform over the cross section;
- the water level across the section is horizontal;
- the streamline curvature is small and the vertical accelerations are negligible, hence the pressure is hydrostatic;
- the effect of boundary friction and turbulence can be accounted for through resistance laws analogous to those used for steady state flow;
- the average bed slope is small, so that the cosine of the angle it makes with the horizontal may be replaced by unity.

A gradually varied flow (wide rectangular channels) is based upon:

$$\frac{dy}{dx} = \frac{S_o - S_f}{1 - Fr^2} \tag{2.37}$$

where:
dy/dx = change of the water depth in the x-direction
dy/dx = slope of the water surface relative to the channel bottom
$\quad S_o$ = bottom slope
$\quad S_f$ = slope of the energy line
$\quad Fr^2$ = Froude number

Depending on whether dy/dx is negative or positive, the following water surface profiles can be distinguished:

$$\frac{dy}{dx} > 0 \implies \text{Backwater curve}$$

$$\frac{dy}{dx} = 0 \implies \text{Uniform flow}$$

$$\frac{dy}{dx} < 0 \implies \text{Drawdown curve}$$

The critical slope for a given discharge Q is by definition that bottom slope for which the normal water depth is equal to the critical depth; $S_o = S_c$ for $y_n = y_c$. Any bottom slope can be compared with this critical slope. Table 2.3 gives a classification of the bottom slopes.

Table 2.3. Type of bottom slope.

Slope description	Bottom slope	Type of slope
Horizontal	$S_o = 0$	H
Mild	$0 < S_o < S_c$	M
Critical	$S_o = S_c$	C
Steep	$S_o > S_c$	S
Adverse or negative	$S_o < 0$	A or N

A particular discharge Q in a canal gives a normal and a critical water depth for a given bottom slope S_o. The actual water depth y in the canal can be related to these two water depths, namely the actual water depth can be either larger or smaller than the two water depths or is between the two water depths (see Table 2.4).

Therefore, the actual water depth y can be classified in one of the following three regions:
Region 1: $y > y_n$ and $y > y_c$
Region 2: y between y_n and y_c
Region 3: $y < y_n$ and $y < y_c$

Table 2.4. Flow types based on the energy line, bottom slope and Froude number.

Actual water depth y in relation to y_n	Relation energy line and bottom slope	Flow type
$y > y_n$	$S_f < S_o$ or $S_o - S_f > 0$	Gradually varied
$y = y_n$	$S_f = S_o$ or $S_o - S_f = 0$	Uniform
$y < y_n$	$S_f > S_o$ or $S_o - S_f < 0$	Gradually varied

Actual water depth y in relation to y_c	Froude number	Flow type
$y > y_c$	$Fr^2 < 1$ or $1 - Fr^2 > 0$	Subcritical
$y = y_c$	$Fr^2 = 1$ or $1 - Fr^2 = 0$	Critical
$y < y_c$	$Fr^2 > 1$ or $1 - Fr^2 > 0$	Supercritical

Another classification of flow types is based on the type of slope and the actual water depth. Table 2.5 gives a summary of the most common water surface profiles in wide canals; a graphical presentation of these profiles is given in Figure 2.12. Figure 2.13 gives an overview of the most common water surface profiles in wide canals. These canals are characterised by a large width B so that the hydraulic radius R is equal to y and the normal and the critical water depth are a function of the specific discharge $q = Q/B$ and the bottom roughness only.

Table 2.5. Summary of water surface profiles.

Bottom slope	Water surface profile 1	2	3	Depth range of y, y_c and y_n Region 1	Region 2	Region 3	Type of curve	Flow type
Steep S	S1			$y > y_c > y_n$			Backwater	Subcritical
$S_o > S_c$		S2			$y_c > y > y_n$		Drawdown	Supercritical
$y_n < y_c$			S3			$y_c > y_n > y$	Backwater	Supercritical
Critical C	C1			$y > y_c = y_n$			Backwater	Subcritical
$S_o = S_c$		C2			$y_c = y = y_n$		Uniform	Critical
$y_n = y_c$			C3			$y < y_c = y_n$	Backwater	Supercritical
Mild M	M1			$y > y_n > y_c$			Backwater	Subcritical
$0 < S_o < S_c$		M2			$y_n > y > y_c$		Drawdown	Subcritical
$y_n > y_c$			M3			$y_n > y_c > y$	Backwater	Supercritical
Horizontal H						n.a.		
$S_o = 0$		H2			$y_n > y > y_c$		Drawdown	Subcritical
$y_n =$ infinite			H3			$y_n > y_c > y$	Backwater	Supercritical
Adverse A						n.a.		
$S_o < 0$		A2			$y > y_n$		Drawdown	Subcritical
$y_n =$ none			A3			$y_n > y$	Backwater	Supercritical

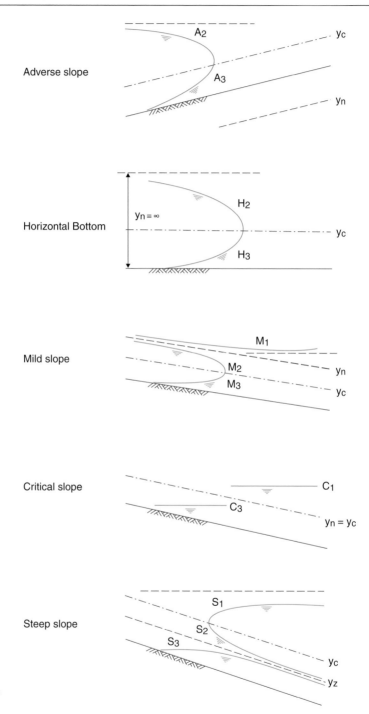

Figure 2.12. Summary of water surface profiles in wide canals ($R = y$ and $q = Q/B$).

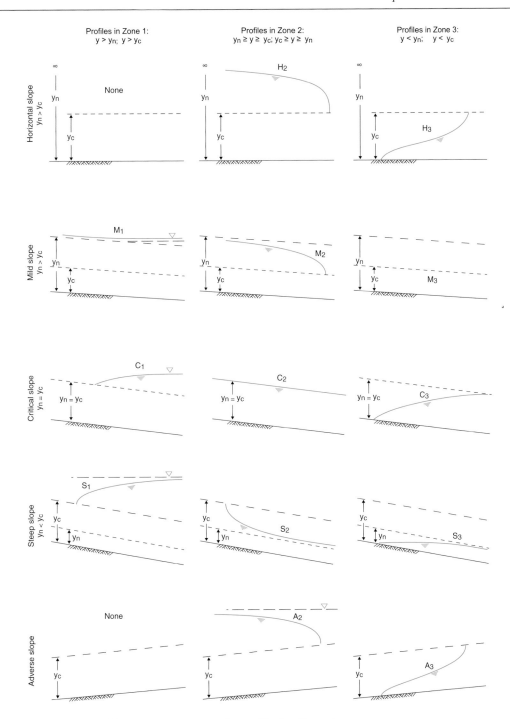

Figure 2.13. Overview of the most common water surface profiles in wide canals.

Gradually varied flow can be computed by three methods: direct integration; graphical integration and numerical integration. Some characteristics of these methods are presented in Table 2.6.

Table 2.6. Summary of the methods for computing gradually varied flow.

Method	Distance from depth	Depth from distance	Cross section	Remarks
Bresse	O	O	Prismatic; broad; rectangle	Use of tables recommended Use of Chézy formula *Recommendation for computation steps*: In upstream direction for subcritical flow In downstream direction for supercritical flow
Bakhmeteff	O		All shapes	Use of tables recommended Use of hydraulic exponents N and M
Graphical integration	O		Prismatic; non-prismatic	For $y \rightarrow y_n$ large errors may occur
Direct step	O		Prismatic	*Recommendation for computation steps:* In upstream direction for subcritical flow In downstream direction for supercritical flow
Standard step		O	All shapes	Iteration required Recommended for natural channels Eddy and other losses can be included Use of field data possible *Recommendation for computation steps*: In upstream direction for subcritical flow; In downstream direction for supercritical flow
Predictor corrector		O	All shapes	Iteration not required Straightforward water depth computation

2.8 SOME GENERAL ASPECTS OF UNSTEADY FLOW

Unsteady flow in open channels occurs when the flow parameters change with time at a fixed point, e.g. the water depth or the discharge varies at a certain point of the canal with time. Waves travelling in canals are examples of unsteadiness; they can be classified as:

• Translatory waves with a net transport of water in the direction of the wave; translatory waves are the most common type of waves in open channels.

• Oscillatory waves – a temporal variation in the water surface that is propagated through water, for instance wind waves, without a net transport of water. Oscillatory waves are normally ignored in the design and operation of irrigation canals.

A translatory wave is a gravity wave with a substantial displacement of water particles in the flow direction. The waves can be divided into gradually and rapidly varied unsteady flow. In gradually varied unsteady flow, the wave profile is gentle and the change in depth is gradual; examples are waves due to a slow gate operation in a canal. Rapidly varied unsteady flow examples are surges caused by the rapid opening or closing of a regulating structure.

A special wave is the solitary wave that is characterised by a rising limb and a single peak, followed and preceded by steady flow. A downstream solitary wave moves down and an upstream wave moves up a channel slope.

A positive wave has an increase in water level and a negative wave has a decrease in water level from a steady flow. A short wave is characterized by a ratio of water depth and wave length smaller than 0.05 ($y/L < 0.05$).

The celerity of a wave is the speed of propagation of the disturbance relative to the water flow. In general, the celerity c for a wave with small amplitude and a two-dimensional flow follows from Airy's theory, which neglects the viscosity and surface tension: $c^2 = (gL/2\pi)\tanh(2\pi y/L)$.

For short waves and small tanh the celerity become: $c^2 = gy$ (the Lagrange equation).

These lecture notes will discuss the sediment transport in irrigation canals. The criteria for the hydraulic design of a canal network will include the requirement that the water level is kept as much as possible at the required supply level and that the amount of water supplied is in line with the water requirements during the growing season. Therefore, it is logical to present some unsteady flow concepts here, especially in view of the sediment transport and water flow as proposed in the next chapters.

As mentioned before, the flow in irrigation networks is usually controlled for the optimal use of the available water, which will result in variations in time. A fast control may even lead to unwanted results, such as the development of hydraulic jumps. Although the design of many irrigation networks is based upon steady flow concepts, the hydraulic performance under actual operation usually requires a closer inspection based on unsteady flow computations.

Unsteady flows are characterized by either more or less significant variations in water depth and discharge, both in time and space. The related hydraulic problems are governed by two concepts, namely storage and conveyance. For an incompressible flow the storage concept deals with the conservation of water mass. The conveyance concept deals with the balance of forces acting upon a water mass and their effect on the momentum balance. In this context, the amount of momentum loss due to channel friction relative to the increase of momentum due to the gravity forces is an essential aspect.

One of the important parameters influencing the capability of a canal to adapt itself to changes is its storage capacity. If the storage capacity in a canal reach is relatively large compared to the difference in inflow and outflow, the time scale of adaptation is also large and the changes in the canal will show a very unsteady flow behaviour. If the storage capacity in a canal reach is small, the adaptation of the canal to the new boundary conditions may be fast and the flow may pass through a series of almost steady state. This adaptation also depends on the capability of the flow to accelerate or decelerate. If the adaptation of the velocity of the flow particles occurs quickly, then the network will pass through a series of nearly steady states; otherwise, the flow in the network will clearly show an unsteady behaviour.

An example is the flow through a structure in an irrigation canal. Usually this flow is assumed to be steady. The discharge through the structure at any moment depends on the water level. Although the water level may vary rapidly, the discharge generated will respond more or less instantaneously. The immediate response of the flow to the changing boundary conditions is due to lack of storage between the upstream and downstream section and the relatively small water mass to be accelerated or decelerated. Generally a hydraulic structure is a steady flow element in a network and how the network as a whole will behave depends on the flow characteristics of the other elements in the network.

2.9 BASIC DIFFERENTIAL EQUATIONS FOR GRADUALLY VARIED UNSTEADY FLOW

Unsteady flow equations will describe the dependent variables, namely the velocity v and water depth y, as function of the independent variables x and t. The definition of the two dependent variables requires the formulation of two equations to solve them. In open channel flow, they are usually based upon the mass (volume) and momentum conservation. The conservation of mass assumes that the fluid is incompressible and that the density is constant. The conservation of mass leads to the continuity equation.

Gradually varied unsteady flow refers to an unsteady flow in which the curvature of the wave profile is mild; the change of depth with time is gradual; the vertical acceleration of the water particles is negligible in comparison with the total acceleration and the effect of the boundary friction can not be neglected. Figure 2.14 gives a schematized wave in the x, t and y direction as an example of unsteady flow.

Only a few gradually varied, unsteady flow problems can be solved analytically; most problems require a numerical solution. The gradually varied unsteady flow can be described by the de St. Venant equations, which consist of the continuity and the dynamic equation. Unsteady flow in open channels is assumed to be a one-dimensional flow with straight and parallel flow lines. The dynamic equation includes the change of velocity v with time and consequently the acceleration, which produces the forces and causes the energy losses in the flow.

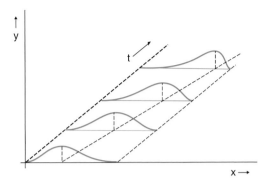

Figure 2.14. An unsteady flow presented in the *x*, *t* and *y* direction.

The dynamic equation

Unsteady flow analysis for open channels deals with the changes of velocity and water depth with position and time. Here the depth *y* and velocity *v* are the dependent variables. Remember that in steady flow the gradient d*E*/d*x* represents the slope of the total energy line and is equal, but opposite in sign to the friction slope $S_f = v^2/C^2R$. When the increase in depth *y* and the downstream distance *x* are taken as positive, the equation for the total energy may be written as:

$$E = z + y + \frac{\alpha v^2}{2g}$$

$$\frac{dE}{dx} = \frac{dz}{dx} + \frac{dy}{dx} + \frac{2v}{2g}\frac{dv}{dx}$$

Remember that $v = f(x,t)$; hence the acceleration can be expressed by partial differentials as:

$$\frac{dv}{dx} = \frac{\partial v}{\partial x} + \frac{\partial v}{\partial t}\frac{\partial t}{\partial x}$$

$$\frac{\partial E}{\partial x} = \frac{\partial z}{\partial x} + \frac{\partial y}{\partial x} + \frac{v}{g}\left(\frac{\partial v}{\partial x} + \frac{\partial v}{\partial t}\frac{\partial t}{\partial x}\right)$$

$$v = \frac{\partial x}{\partial t} \quad \text{or} \quad \frac{\partial t}{\delta x} = \frac{1}{v}$$

$$-S_f = -S_o + \frac{\partial y}{\partial x} + \frac{v}{g}\frac{\partial v}{\partial x} + \frac{1}{g}\frac{\partial v}{\partial t}$$

Thus:

$$S_f = S_o - \frac{\partial y}{\partial x} - \frac{v}{g}\frac{\partial v}{\partial x} - \frac{1}{g}\frac{\partial v}{\partial t} \qquad \text{(de Saint Venant equation)} \qquad (2.38)$$

$$S_f = S_o - \frac{\partial y}{\partial x} - \frac{v}{g}\frac{\partial v}{\partial x} - \frac{1}{g}\frac{\partial v}{\partial t} = \frac{v^2}{C^2 R} \qquad (2.39)$$

This is the de Saint Venant equation, where:
y = water depth (m)
t = time (s)
v = average velocity (m/s)
x = longitudinal distance (m)
S_o = channel slope (m/m)
S_f = friction slope = slope of the energy line (m/m)

By definition $Q = vA$ and $v = Q/A$

$$S_o - S_f - \frac{\partial y}{\partial x} - \frac{v}{gA}\frac{\partial Q}{\partial x} - \frac{1}{gA}\frac{\partial Q}{\partial t} = 0 \qquad (2.40)$$

Dynamic equation for gradually varied unsteady flow
This general dynamic equation for gradually varied unsteady flow is only true when the pressure is hydrostatic, meaning that the vertical component of the acceleration is negligible and that the flow lines are straight and parallel.

The first two terms on the right hand side ($S_o - S_f = 0$) signifies steady uniform flow. The first four terms together [$S_o - S_f - \partial y/\partial x - v/gA(\partial Q/\partial x) = 0$] express a steady gradually varied flow.
The dynamic equation may also be written in another form:

$$\frac{\partial y}{\partial x} + \frac{v}{g}\frac{\partial v}{\partial x} + \frac{1}{g}\frac{\partial v}{\partial t} + S_f - S_o = 0 \qquad (2.41)$$

$$g\frac{\partial y}{\partial x} + v\frac{\partial v}{\partial x} + \frac{\partial v}{\partial t} + g(S_f - S_o) = 0 \qquad (2.42)$$

These equations clearly show the four variables in unsteady, open channel flow, namely x, t, v and y.

Continuity equation
The law of continuity for an unsteady, one-dimensional flow can be found by considering the conservation of mass of a small canal reach (dx) between two cross-sections. The difference in outflow and inflow in the reach during a time step dt is equal to the change of storage over that

distance dx. Moreover, assume that the surface width B_s is constant (prismatic section) and that the water is incompressible. For the unsteady flow conditions the discharge Q changes with distance dx, namely dQ/dx and the water depth y changes with time: dy/dt. The change of discharge Q through space in a small time-step dt is (dQ/dx)dx dt. The corresponding change in storage in space is B_s dy/dt dx dt. From continuity considerations (volumetric balance) follows that a change of discharge Q in Δx-direction must be accompanied by a change in water depth y in time step Δt:

$$\left(\frac{\mathrm{d}Q}{\mathrm{d}x}\right)\mathrm{d}x\,\mathrm{d}t + B_s\frac{\mathrm{d}y}{\mathrm{d}t}\mathrm{d}x\,\mathrm{d}t = 0$$

$$\frac{\partial Q}{\partial x} + B_s\frac{\partial y}{\partial t} = 0 \qquad \textit{Continuity equation for unsteady flow} \qquad (2.43)$$

From this equation follows that the change in discharge Q in the Δx-direction is opposite to the change in water depth y during time step Δt.

Assuming that there is no lateral discharge (no inflow or outflow: $\Delta q = 0$) in Δx and with $Q = vA$

$$\frac{\partial Q}{\partial x} = \frac{\partial A\cdot v}{\partial x} = v\frac{\partial A}{\partial x} + A\frac{\partial v}{\partial x}$$

$$A\frac{\partial v}{\partial x} + v\frac{\partial A}{\partial x} + B_s\frac{\partial y}{\partial t} = 0 \qquad \begin{array}{l}\textit{Continuity equation for a channel}\\ \textit{with a general shape}\end{array} \qquad (2.44)$$

The first term on the left-hand side represents the prism storage, the second term represents the wedge storage and $B_s\,\partial y/\partial t$ is the rate of rise of the water level with time t.

Prism storage + wedge storage + rate of rise of water level = 0

With the hydraulic depth $D = A/B_s$ and with d$A = B_s$ dy, the continuity equation becomes:

$$D\frac{\partial v}{\partial x} + v\frac{\partial y}{\partial x} + \frac{\partial y}{\partial t} = 0 \qquad \begin{array}{l}\textit{Continuity equation for a channel}\\ \textit{with a general shape}\end{array} \qquad (2.45)$$

For a rectangular channel with $q = Q/B$, $v = q/y$ and $A = B_s y$ follows:

$$y\frac{\partial v}{\partial x} + v\frac{\partial y}{\partial x} + \frac{\partial y}{\partial t} = 0 \qquad \begin{array}{l}\textit{Continuity equation for a}\\ \textit{rectangular channel}\end{array} \qquad (2.46)$$

where:
y = water depth (m)
t = time (s)
v = average velocity of the flow (m/s)
x = longitudinal distance (m)

Summarising, the continuity and dynamic equations for unsteady flow in open channels read:

$$v\frac{\partial A}{\partial x} + A\frac{\partial v}{\partial x} + B_{\mathrm{s}}\frac{\partial y}{\partial t} = 0 \quad \begin{array}{l} \textit{Continuity equation for unsteady} \\ \textit{flow for a channel with general shape} \end{array} \quad (2.47)$$

$$\frac{\partial y}{\partial x} + \frac{v}{g}\frac{\partial v}{\partial x} + \frac{1}{g}\frac{\partial v}{\partial t} + S_{\mathrm{f}} - S_{\mathrm{o}} = 0 \quad \begin{array}{l} \textit{Dynamic equation for} \\ \textit{unsteady flow} \end{array} \quad (2.48)$$

The continuity and dynamic equation in this form are also known as the de St. Venant equations.

From a combination of the dynamic equation and the continuity equation follows:

$$\frac{\partial Q}{\partial t} - \frac{2QB_{\mathrm{s}}}{A}\frac{\partial y}{\partial t} + gA\left(1 - \frac{Q^2 B_{\mathrm{s}}}{gA^3}\right)\frac{\partial y}{\partial x} - gAS_{\mathrm{o}} + gAS_{\mathrm{f}} = 0 \quad (2.49)$$

Combination of the dynamic and continuity equation for a canal with a general shape.

2.10 SOLUTION OF THE DE ST. VENANT EQUATIONS

The continuity and dynamic equation compose a set of gradually varied, unsteady flow equations, which together form a complete dynamic model of the flow. This complete model can provide accurate results for an unsteady flow, but at the same time, the model can be very demanding in view of the required computations. Moreover, the model is limited by the assumptions made in the deduction of the de St. Venant equations and the suppositions required for their application for specific problems, e.g. assumptions regarding channel irregularities. The set of the two simultaneous equations has to be solved for the two unknowns, v and y, given appropriate boundary conditions and initial conditions.

At present three main numerical methods are available to solve the de St. Venant equations, namely:
- Finite differences (FD)
- Method of characteristics (MOC)
- Finite element method (FEM)

The solution of the de St. Venant equations is complicated, also for simple rectangular channels. General solutions are only possible by using one of the numerical methods (see Table 2.7). The method with finite differences is the most common one. The method of characteristics is adequate for waves due to sudden gate operations. The finite element method can accommodate canals with irregular boundaries.

Table 2.7. Summary of the methods to solve the de St. Venant equations.

de St. Venant Equations				
Numerical Methods				
Direct		Method of characteristics		Finite element method	Approximate Methods
Implicit	Explicit	Characteristic nodes Implicit Explicit	Rectangular grid Implicit Explicit		Storage routing Muskingum Diffusion analogy Kinematic wave

The solution method of the various numerical procedures can be explicit or implicit:
- Explicit: the discharge Q and water depth y at the next time-step are expressed in terms of the discharge Q and water depth y at the current time step; this method is straightforward, but it must be in line with the Courant stability criterion ($c(\Delta t/\Delta x) < 1$). This criterion imposes a severe limitation on the time-distance grid.
- Implicit: the discharge Q and water depth y are related to subsequent and previous time steps, which is a more complex solution procedure, but a stable one.

2.11 RECTANGULAR CHANNELS AND THE METHOD OF CHARACTERISTICS

The curvature of the wave front of a gradually varied unsteady flow is mild and the change in water depth is gradual. The vertical component of the acceleration is negligible in comparison to the total acceleration; the effect of the friction is significant and should not be neglected in the unsteady flow analysis. In irrigation systems the occurrence of rapid varied flow should be prevented by a gradual operation of the regulating and control structures.

One of the main advantages of the method of characteristics is the possibility to visualize the way in which flow disturbances or the effects of flow control move through a canal network. The structure of the net of characteristics will improve the understanding of the numerical procedures required for the practical solution of hydraulic phenomena in canal system. Therefore a short explanation of the method of characteristics will be given here; hydraulic reference books present more details on unsteady flow analysis.

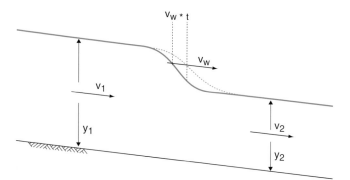

Figure 2.15. Propagation of a wave front in downstream direction.

The set of continuity and momentum equations is very useful for the formulation of solutions on the basis of finite differences, but a transformation of the equations might help to understand the continuity and dynamic equations more easily. For the transformation the auxiliary variable $c = \sqrt{gD}$ (Lagrange wave speed) will be used.

The transformation will use the de St. Venant equations and the differential of the area A to x:

$$\frac{\partial A}{\partial x} = \frac{\partial (\alpha B_s y)}{\partial x} = B_s \frac{\partial y}{\partial x} + \alpha y \frac{\partial B_s}{\partial x}$$

The term $\alpha y (\partial B_s / \partial x)$ accounts for non-prismatic channels and $\partial B_s / \partial x$ is the change of the surface width in the flow direction. The coefficient α is the Coriolis coefficient. The value of α depends on the geometry of the cross-section. The value of α is about 0.5 for triangular and about 1.0 for rectangular cross-sections.

The continuity equation can be written as:

$$v B_s \frac{\partial y}{\partial x} + \alpha y v \frac{\partial B_s}{\partial x} + A \frac{\partial v}{\partial x} + B_s \frac{\partial y}{\partial t} = 0$$

$$v \frac{B_s}{A} \frac{\partial y}{\partial x} + \frac{\alpha y v}{A} \frac{\partial B_s}{\partial x} + \frac{\partial v}{\partial x} + \frac{B_s}{A} \frac{\partial y}{\partial t} = 0$$

By definition the hydraulic depth D is A/B_s

$$\frac{v}{D} \frac{\partial y}{\partial x} + \frac{\partial v}{\partial x} + \frac{1}{D} \frac{\partial y}{\partial t} + \frac{\alpha y v}{A} \frac{\partial B_s}{\partial x} = 0$$

$$v\frac{\partial y}{\partial x} + D\frac{\partial v}{\partial x} + \frac{\partial y}{\partial t} + \frac{\alpha y v D}{A}\frac{\partial B_s}{\partial x} = 0$$

Rearrange this equation and introduce the auxiliary variable $c = \sqrt{gD}$ (Lagrange), where c is the velocity of a long wave in water depth y. In other words c becomes a measure of y.

$$v\frac{\partial y}{\partial x} + \frac{c^2}{g}\frac{\partial v}{\partial x} + \frac{\partial y}{\partial t} + \frac{\alpha v y}{A}D\frac{\partial B_s}{\partial x} = 0$$

For a rectangular channel the surface width B_s and bottom width B are equal and therefore $dB_s/dx = 0$ and the hydraulic depth become $D = A/B_s = y$. Substituting the Lagrange wave speed approximation, namely $gy = c^2$; $dy = 2c/g\, dc$ and $dy/dx = 2c/g\, dc/dx$, the dynamic equation $S_f = S_o - (\partial y/\partial x) - (v/g)(\partial v/\partial x) - (1/g)(\partial v/\partial t)$ will change into:

$$2c\frac{\partial c}{\partial x} + v\frac{\partial v}{\partial x} + \frac{\partial v}{\partial t} = g(S_o - S_f) \qquad \begin{array}{l}\textit{Dynamic equation for}\\ \textit{a rectangular channel}\end{array} \qquad (2.50)$$

Substituting the Lagrange wave speed approximation in the continuity equation for a rectangular channel, which is $y(\partial v/\partial x) + v(\partial y/\partial x) + (\partial y/\partial t) = 0$, results in:

$$2v\frac{\partial c}{\partial x} + 2\frac{\partial c}{\partial t} + c\frac{\partial v}{\partial x} = 0 \qquad \begin{array}{l}\textit{Continuity equation for}\\ \textit{a rectangular channel}\end{array} \qquad (2.51)$$

By writing first the sum and then the difference of the two above given equations, two new equations will be obtained. First, the addition will be presented.

1. The sum of the continuity equation and the dynamic equation results in:

$$2v\frac{\partial c}{\partial x} + c\frac{\partial v}{\partial x} + 2\frac{\partial c}{\partial t} + 2c\frac{\partial c}{\partial x} + v\frac{\partial v}{\partial x} + \frac{\partial v}{\partial t} + g(S_f - S_o) = 0$$

$$2(v + c)\frac{\partial c}{\partial x} + 2\frac{\partial c}{\partial t} + (v + c)\frac{\partial v}{\partial x} + \frac{\partial v}{\partial t} + g(S_f - S_o) = 0$$

$$(v + c)\frac{\partial(v + 2c)}{\partial x} + \frac{\partial(v + 2c)}{\partial t} = g(S_o - S_f) \qquad (2.52)$$

2. Similarly the continuity equation minus the dynamic equation:

$$(v - c)\frac{\partial(v - 2c)}{\partial x} + \frac{\partial(v - 2c)}{\partial t} = g(S_o - S_f) \qquad (2.53)$$

Finally a combination of the two equations results in a general equation for non-steady flow:

$$(v \pm c)\frac{\partial(v \pm 2c)}{\partial x} + \frac{\partial(v \pm 2c)}{\partial t} - g(S_\mathrm{o} - S_\mathrm{f}) = 0 \qquad \text{\textit{General equation of non-steady flow in a rectangular channel}}$$

$$(2.54)$$

Remember that the general mathematical expression for a change of a function f in the x-t-diagram when going from a given point 1 to a neighbouring point 2 can be given by the partial derivatives of the function in the t-direction and the x-direction.

$$df = \frac{\partial f}{\partial t}dt + \frac{\partial f}{\partial x}dx \qquad \text{or} \qquad \frac{df}{dt} = \frac{\partial f}{\partial x}\frac{dx}{dt} + \frac{\partial f}{\partial t}$$

In this equation f is a variable dependent on the two independent variables x and t, and the equations give the rate of change of f if x and t are simultaneously varied in a prescribed manner, given by dx/dt. Assume that the function f is, for example, the water depth y ($f = y(x, t)$).

$$\frac{dy}{dt} = \frac{\partial y}{\partial x}\frac{dx}{dt} + \frac{\partial y}{\partial t}$$

Assuming that the function y is the water depth then the situation can be considered in the following way: to an observer walking with a speed dx/dt along a dike of an open canal, the depth y will appear to vary with time at the rate given by the equation for dy/dt. A similar result would of course be true for any other parameter such as v, q, or c.

The total differential du/dt for the function $u = v + 2c$ can be found in the same way as dy/dt and will give:

$$\frac{du}{dt} = \frac{\partial u}{\partial x}\frac{dx}{dt} + \frac{\partial u}{\partial t} \qquad \text{with } v = f(x, t) \text{ and } \frac{dx}{dt} = v + c$$

Assume that S_o is constant and that S_o and S_f are both relatively small. Then from this assumption follows that $g(S_\mathrm{o} - S_\mathrm{f})$ is also very small and a new set of equations for the unsteady flow can be written as:

$$(v \pm c)\frac{\partial(v \pm 2c)}{\partial x} + \frac{\partial(v \pm 2c)}{\partial t} = g(S_\mathrm{o} - S_\mathrm{f}) \qquad (2.55)$$

$$(v \pm c)\frac{\partial(v \pm 2c)}{\partial x} + \frac{\partial(v \pm 2c)}{\partial t} = 0 \qquad \text{\textit{Equation of non-steady flow in a rectangular channel}} \qquad (2.56)$$

The first two terms at the left-hand side, namely $(v \pm c)(\partial(v \pm 2c)/\partial x) + \partial(v \pm 2c)/\partial t = g(S_o - S_f)$, represents the rate of change of $(v \pm 2c)$ from the view point of two observers, one moving in the $x-t$ plane with velocity $(v + c)$ and the other moving with a velocity $(v - c)$.

The path of the two imaginary observers can be plotted on the $x-t$ plane and a complete solution will be obtained for any prescribed unsteady flow situation. Only in simple cases does the process lead to explicit solutions, but in more complex cases numerical methods may be used without any great difficulty (see Figure 2.16).

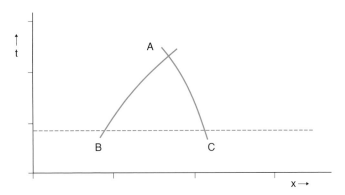

Figure 2.16. Method of characteristics in the x and t plane, with an explicit solution.

Along the line with the direction $dx/dt = (v \pm c)$, the expression $dx/dt = (v \pm c)$ holds and along the line with the direction $dx/dt = v - c$ the expression $dx/dt = (v \pm c)$ holds. The line with the direction $dx/dt = (v \pm c)$ is called the positive characteristic (c_+) and the other is the negative characteristic (c_-).

Integration of $dx/dt = (v \pm c)$ gives that $v + 2c = $ constant. The equation $dx/dt = (v \pm c)$ means that $v \pm c$ can be treated as constant for Δt. If $dx/dt = (v \pm c)$ then $u = v \pm 2c + g(S_f - S_o)t$. The reciprocal value gives $dt/dx = 1/v \pm c$ which defines two lines in $x-t$ space.

Thus in this specific case that two observers move with two velocities, namely $(v \pm 2c)$, the two velocities $(v \pm 2c)$ appear to remain constant. The results are two families of curves in the $x-t$ plane, which have inverse slopes, namely $(v + c)$ and $(v - c)$. These lines are the characteristic lines.

Summary of the four characteristic equations of unsteady flow in a rectangular channel
The method of characteristics gives two lines:
- one line has a slope $1/(v + c)$
- the other line has a slope $1/(v - c)$

Along these two lines $\dfrac{du}{dt} = \dfrac{\partial u}{\partial x}\dfrac{dx}{dt} + \dfrac{\partial u}{\partial t}$ and $\dfrac{du}{dt} = 0$

$\dfrac{dt}{dx} = \dfrac{1}{v+c}$ defines a positive characteristic along which $v + 2c$ is constant

$\dfrac{dt}{dx} = \dfrac{1}{v-c}$ defines a negative characteristic along which $v - 2c$ is constant

In principle the terms $v \pm 2c$ and $g(S_f - S_o)$ are constant

$$\frac{dt}{dx} = \frac{1}{v \pm c} \tag{2.57}$$

$$\frac{d(v \pm 2c + g(S_f - S_o)t)}{dt} = 0 \tag{2.58}$$

The solution of the method of characteristics is based on the construction of a network of points in the $x-t$ plane where the dependent variables are computed. From a given set of two points on the network, a new point may be constructed by drawing a characteristic of one family from one point that intersects with the characteristics of the other family drawn from the other point. In general, these directions are not known initially as

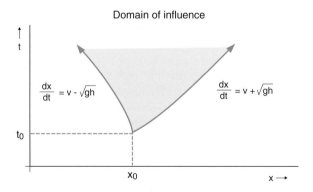

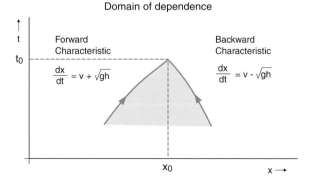

Figure 2.17. Characteristics in the x and t plane with an example of the domain of influence and of the domain of dependence.

they depend on the solution of the dependent variables at the intersection point (see Figure 2.17). However, the solution can be found through the Riemann invariants. The almost exactdirection of these characteristics can then be drawn based on the average characteristic direction at the connecting points.

The significance of the method of characteristics lies in the fact that the characteristics are to be seen as lines along which some information on the state of the fluid propagates. Along the characteristic lines, a special condition is valid, which implies that these lines are unique lines and that information only travels along these characteristics. With the characteristics, it is possible to compute solutions for v and y from known conditions at an earlier point in time.

In the derivation of the method of characteristics, the auxiliary variable $c = \sqrt{gD}$ has a special meaning as part of the direction of the characteristics. For example when $v = 0$, the direction of both characteristics is the same in magnitude, but they have a different, opposite sign. This means that the effect of any disturbance at a certain point in a canal propagates in both directions (upstream and downstream) with the same speed c. The variable c here means the celerity with which disturbances propagate in stagnant water.

When $v > 0$, the characteristic directions no longer have the same magnitude and at a particular point information on the fluid state travels faster in the downstream than in the upstream direction. For smaller velocities the flow is subcritical (the Froude number is less than unity) and the state of flow at any point in time is controlled by both the upstream and downstream conditions.

CHAPTER 3

Sediment Properties

3.1 INTRODUCTION

Sediments are fragmented material, primarily formed by the physical and chemical disintegration of rocks from the earth's crust; they can be divided into cohesive and non-cohesive sediments. For non-cohesive sediments there are no physical-chemical interactions between individual particles and the size and weight of the individual particles are important factors in their behaviour. In cohesive sediments the physical-chemical interactions between particles are important factors in the initiation of motion (erosion) and also in the transportation (flocculation); with cohesive sediments the size and weight of a particle have less significance. Most of the discussion on sediment transport in these lecture notes will deal with non-cohesive sediments.

Some of the major features of the individual particles and the sediments can be described by:
- Density and porosity
- Particle size and size distribution
- Shape
- Fall velocity
- Dimensionless parameters, such as the particle parameter, particle mobility parameter, excess shear stress parameter, dimensionless particle Reynolds number and transport rate parameter.

3.2 DENSITY AND POROSITY

The density of a sediment particle is the mass per unit of volume and it primarily depends on the mineral composition. Non-cohesive material originates generally from the disintegration or decomposition of quartz. The density (ρ_s) of quartz is approximately 2650 kg/m^3 and the density of clay minerals ranges from 2500–2700 kg/m^3.

The relative density is often expressed as Δ.

$$\Delta = \frac{(\rho_s - \rho_w)}{\rho_w} \tag{3.1}$$

The specific density s is expressed as:

$$s = \frac{\rho_s}{\rho_w} \tag{3.2}$$

$\Delta = s - 1 =$ relative density

where:

$\Delta =$ relative density; the relative density of quartz in water equals 1.65

$s =$ specific density; the specific density of quartz in water equals 2.65

$\rho_s =$ density of the sediment particle (kg/m^3)

$\rho_w =$ density of water (kg/m^3)

Table 3.1 gives the density of water as a function of the temperature.

Table 3.1. The density of fresh water as a function of the temperature T.

T (°C)	0	4	12	16	21	32
ρ_w (kg/m^3)	999.87	1000.0	999.5	999.0	998.0	995.0

The dry bulk density of sediment is the mass per unit of volume on a dry weight basis and includes the pores between the particles. The bulk density is equal to:

$$\rho_{dry} = (1 - p)\rho_s \tag{3.3}$$

where:

$p =$ porosity (dimensionless)

$p =$ the ratio of the volume pores and the total volume of a sediment sample

$\rho_s =$ density of the sediment particles (kg/m^3)

$\rho_{dry} =$ dry bulk density of the sediment (kg/m^3)

The porosity p often depends on the deposition and erosion processes. The dry bulk density of sand (ρ_{s-dry}) is approximately 1500–1550 kg/m^3. For silt the bulk density ranges between 1050 and 1320 kg/m^3, while for clay these value may range between 500 and 1250 kg/m^3.

Sometimes the specific weight is still used and is expressed as $\gamma = \rho_s g$ with γ in Newtons per unit of volume (N/m^3).

3.3 SIZE AND SIZE DISTRIBUTION

A first attempt to classify the individual, non-cohesive sediment particles, is based on their size; more specifically their diameter. Usually, sediments are referred to as gravel, sand, silt or clay, which refer to the size of the individual particles; listed here in decreasing magnitude. Many attempts have been made to describe the sediment by one distinct diameter and a range of sizes has been suggested, but none of them has proven to be entirely suitable. Some of the main definitions for the diameter (Bogardi, 1974) include:

- Sieve diameter (d): diameter of the square mesh sieve, which will just let the particle pass. The sieve diameter is generally used for fractions greater than 0.1 mm.
- Settling or sedimentation diameter (d_s): diameter of a sphere with the same density and same fall velocity as the given particle in the same fluid and at the same temperature.
- Nominal diameter (d_n): diameter of a sphere or cube with the same volume as the particle.
- Tri-axial dimensions (a, b, c): the dimensions of a particle in the direction of three orthogonal axes. The largest dimension is called 'a' and the smallest dimension is called 'c'.

The sizes can be determined by direct measurements, by sieving, by sedimentation or by microscope analysis. Table 3.2 shows a classification of sediments as recommended by the American Geophysical Union.

Table 3.2. Classification of sediments according to their size.

Classification	mm $= 10^{-3}$ m	μm $= 10^{-6}$ m
Very coarse gravel	64–32	
Coarse gravel	32–16	
Medium gravel	16–8	
Fine gravel	8–4	
Very fine gravel	4–2	
Very coarse sand	2–1	2000–1000
Coarse sand	1.00–0.50	1000–500
Medium sand	0.50–0.25	500–250
Fine sand	0.25–0.125	250–125
Very fine sand	0.125–0.062	125–62
Coarse silt	0.062–0.031	62–31
Medium silt	0.031–0.016	31–16
Fine silt	0.016–0.008	16–8
Very fine silt	0.008–0.004	8–4
Coarse clay	0.004–0.002	4–2
Medium clay	0.002–0.001	2–1
Fine clay	0.001–0.0005	1–0.5
Very fine clay	0.0005–0.00024	0.5–0.24

Sediments normally have many particles with different sizes. The particle sizes result in a distribution, which is generally expressed as percent by mass (weight) versus particle size. The most common method to determine the distribution of particle sizes (size frequency) is a sieve analysis, which can be used for particles larger than 74 μm. Next, the results are presented as a cumulative size-frequency curve. The fraction or the percentage of the sediment (by mass) that is smaller than a given size is plotted against the particle size (sometimes a fraction that is larger than a given size is plotted). The cumulative size distribution of various sediment sizes by mass can be approximated by a log-normal distribution, as shown in Figure 3.1. A log-normal distribution might result in a straight line when logarithmic probability paper is used.

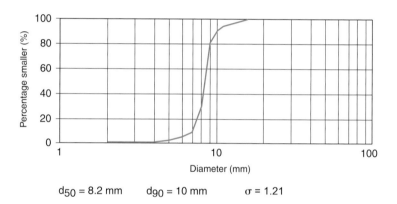

Figure 3.1. Example of a particle size distribution.

d_{50} = 8.2 mm d_{90} = 10 mm   σ = 1.21

From this cumulative size distribution the following sediment characteristics can be defined (van Rijn, 1993):
(a) Mean diameter (d_m or $\overline{d}$) is defined as:

$$d_m = \overline{d} = \frac{\sum P_i d_i}{\sum P_i} \tag{3.4}$$

where:
P_i = fraction (percentage) with diameter d_i (%)
d_i = geometric mean of the size fraction limits = diameter for which i % of the sample is finer than d_i (mm)

(b) The median diameter (d_{50}) is assumed to give the best representation of the sediment mixture and is described by the diameter 'd' for which 50% of the sample is finer than 'd'. Sometimes other values than d_{50} are used in the sediment transport predictors; examples are d_{16}, d_{35}, d_{65}, d_{84}
(c) The geometric mean diameter (d_g) is defined as:

$$d_g = \sqrt{d_{84}\, d_{16}} \tag{3.5}$$

where:

d_{84} = diameter for which 84% of the sample is finer than d_{84}

d_{16} = diameter for which 16% of the sample is finer than d_{16}

(d) The standard deviation (σ_s) is described by:

$$\sigma_s = 0.5\left(\frac{d_{84}}{d_{50}} + \frac{d_{50}}{d_{16}}\right) \tag{3.6}$$

where:

d_{84} = diameter for which 84% of the sample is finer than d_{84}

d_{16} = diameter for which 16% of the sample is finer than d_{16}

d_{50} = diameter d for which 50% of the sample is finer than d_{50}

(e) The geometric standard deviation (σ_g) is given by:

$$\sigma_g = \sqrt{\frac{d_{84}}{d_{16}}} \tag{3.7}$$

where:

d_{84} = diameter for which 84% of the sample is finer than d_{84}

d_{16} = diameter for which 16% of the sample is finer than d_{16}

3.4 SHAPE

In addition to the diameter the shape of the sediment particle is also an important aspect. A flat particle will have a smaller fall velocity than a rounded particle and it will be more difficult to transport a flat particle as bed load. The shape of a particle can be characterized by sphericity, roundness or shape factor.

- Sphericity is the ratio between the surface area of a sphere and the surface area of a particle at equal volume.
- Roundness is the ratio of the average radius of the curvature of the edges and the radius of a circle inscribed in the maximum-projected area of the particle.

Sphericity and roundness are not used in sediment theories.

The shape factor is a useful definition that can be used to describe a particle. The shape factor (s.f.) is defined by:

$$\text{s.f.} = \frac{c}{\sqrt{ab}} \tag{3.8}$$

Here a, b and c are the dimensions of a particle in the direction of three axes. For spheres the shape factor is 1. For natural sand the shape factor (s.f.) is approximately 0.7 and for gravel it is approximately 0.9.

3.5 FALL VELOCITY

The velocity of sediment particles settling in a liquid is called the fall velocity w_s. This velocity is an important physical value to describe the sedimentation and suspension behaviour of sediment particles. The basic fall velocity of sediment particles is derived from a single sphere with diameter d falling with a constant velocity in quiescent (quiet) water. At the beginning of the settling process, the fall velocity is small and the force of gravity is greater than the resistance. Hence, the sphere moves with acceleration, and the resistance increases with the velocity. In time the resistance equals the force of gravity and the sphere then falls with a constant velocity. The force of gravity acting on the sphere is in equilibrium with the resistance to the motion, which depends on the velocity and a drag coefficient.

From resistance force = gravity force follows:

$$C_d \frac{1}{2} \rho_w w_s^2 \frac{\pi}{4} d^2 = \frac{\pi}{6} d^3 g(\rho_s - \rho_w) \tag{3.9}$$

$$w_s = \sqrt{\frac{4}{3} \frac{gd\Delta}{C_d}} \tag{3.10}$$

The drag coefficient C_d is a function of the Reynolds number ($w_s\,d/v$). For low Reynolds numbers $C_d = 24/\mathrm{Re}$.

For natural sand, a Reynolds number lower than 0.1 and $C_d = 24/\mathrm{Re}$, Stokes law gives the fall velocity as:

$$w_s = \frac{(s-1)gd^2}{18v} \quad 1 \le d \le 100\,\mu\mathrm{m} \tag{3.11}$$

where:
w_s = fall velocity in clear water (m/s)
s = specific density of the sediment particle (dimensionless)
g = acceleration due to gravity (m/s^2)
d = diameter of the particle (m)
v = kinematic viscosity (m^2/s)

During the settling process, the motion of a particle also causes the surrounding fluid to move. If the Reynolds number ($\mathrm{Re} = w_s\,d/v$) is less than approximately 0.4, the effect of inertia forces induced in the fluid by the motion is much less than those due to the fluid viscosity. The settling of a single particle affects a large region of the surrounding fluid if the Reynolds number is low. When the inertia forces in the fluid are negligible then the drag coefficient is inversely proportional to the Reynolds number. The fall velocity is proportional to the square of the sphere diameter and the density, but it is inversely proportional to the viscosity of the fluid. Stokes law is valid for Re < 0.4 and in water at normal temperature the fall velocity is valid for particles smaller than $d = 0.076$ mm.

For larger quartz particles Stokes law has to be adjusted and the fall velocity reads:

$$w_s = \frac{10\nu}{d} \left[\left(1 + \frac{0.01(s-1)gd^3}{\nu^2} \right)^{0.5} - 1 \right] \quad 100 \leq d \leq 1000\,\mu m$$

(3.12)

$$w_s = 1.1\sqrt{(s-1)gd} \quad d \geq 1000\,\mu m \tag{3.13}$$

where:
w_s = fall velocity in clear water (m/s) (see Table 3.3)
s = relative density of the sediment particles (dimensionless)
g = acceleration due to gravity (m/s²)
d = diameter (m)
ν = kinematic viscosity (m²/s)

Table 3.3. Fall velocity for sediment particles.

d (mm)	0.06	0.1	0.2	0.3	0.4	0.5	0.75	1.0
w_s (mm/s)	3.2	8.9	25	44	59	72	98	140

The presence of a large number of other particles during the settling process will decrease the fall velocity of a single particle. The fall velocity of a single particle influenced by the presence of other particles follows from:

$$w_{s,m} = (1-c)^\gamma w_s \quad \text{with } 2.3 \leq \gamma \leq 4.6 \quad \text{and} \quad 0 \leq c \leq 0.3 \tag{3.14}$$

where:
$w_{s,m}$ = fall velocity influenced by other particles (m/s)
w_s = fall velocity in clear water (m/s)
c = volumetric sediment concentration (percent)
γ = coefficient, which is a function of the Reynolds number
    (see Table 3.4)

The coefficient γ is slightly dependent on the particle shape, but this can be ignored. For fine sediments with a concentration of 1%, the reduction in fall velocity will be around 5%. The fall velocity of a particle in turbulent water is different from the velocity in quiescent water. A cluster of particles (cohesive sediments) will have a greater fall velocity.

Table 3.4. Coefficient γ as function of Reynolds number.

Reynolds number	<0.2	0.2–1	1–200	>200
γ	4.65	$4.35\,\mathrm{Re}^{-0.03}$	$4.45\,\mathrm{Re}^{-0.1}$	2.39

3.6 CHARACTERISTIC DIMENSIONLESS PARAMETERS

The equations of motion and continuity, both for water and sediment, roughly define the transport of water and sediment. An analytical description of all the physical processes by the equations is not yet possible and therefore, the sediment transport theories still rely on data from field and experimental investigations and from dimensional analysis. Based on dimensional analysis several processes related to sediment transport can be expressed as a function of independent dimensionless parameters. The following parameters are widely used to describe the sediment transport in open channels (van Rijn, 1993).

Particle parameter (D_*) reflects the influence of gravity, density and viscosity on sediment transport and is given by:

$$D_* = \left[\frac{(s-1)g}{v^2} \right]^{1/3} d_{50} \tag{3.15}$$

where:

s = specific density of the sediment particle (dimensionless)
g = acceleration due to gravity (m/s^2)
d_{50} = median diameter (m)
v = kinematic viscosity (m^2/s)

Particle mobility parameter (θ) is the ratio of the drag force and the weight of the submerged particle and reads:

$$\theta_{cr} = \frac{u_{*cr}^2}{(s-1)gd_{50}} = \frac{\tau_{cr}}{(s-1)\rho g d_{50}} \tag{3.16}$$

where:

τ = shear stress (N/m^2)
τ_{cr} = critical shear stress according to Shields (N/m^2)
u_* = shear velocity (m/s)
$u_* = \sqrt{(\tau/\rho)}$
u_{*cr} = critical shear velocity (m/s)
$u_{*crit} = \sqrt{(\tau_{crit}/\rho)}$
ρ = density (kg/m^3)
g = acceleration due to gravity (m/s^2)

d_{50} = median diameter (m)

s = specific density of the sediment particle (dimensionless)

Excess bed shear stress parameter (T) is defined as:

$$T = \frac{\tau' - \tau_{cr}}{\tau_{cr}} \tag{3.17}$$

where:

τ_{cr} = critical shear stress according to Shields (N/m^2)

τ' = grain shear stress (N/m^2), which is the skin resistance or the surface drag due to the grain roughness (see Chapter 5 for more details)

The particle Reynolds number parameter (Re$_*$) is represented by

$$\text{Re}_* = \frac{u_* d_{50}}{\nu} \tag{3.18}$$

where:

u_* = shear velocity (m/s)

d_{50} = median diameter (m)

ν = kinematic viscosity (m^2/s)

Transport rate parameter (ϕ) is represented by:

$$\phi = \frac{q_s}{(s-1)^{0.5} g^{0.5} d_{50}^{1.5}} \tag{3.19}$$

where:

q_s = volumetric sediment transport rate (m^3/s per m width)

s = specific density of sediment (dimensionless)

g = acceleration due to gravity (m/s^2)

d_{50} = median diameter (m)

The concentration (C) is the ratio between the sediment transport rate q_s and the discharge q per m width (both in volume). Here, q_s is the volume of sediment transported in 1 m^3 of water-sediment mixture. The volumetric concentration reads as:

$$C = \frac{q_s}{q} \tag{3.20}$$

where:

C = concentration of sediment in the water by volume (percentage or ppm)

q_s = volumetric sediment transport rate (m^3/s per m width)

q = discharge (m^3/s per m width)

ppm = parts per million

The concentration C can be also expressed as a percentage by mass (C_g), which is the mass of sediment transported in a unit volume of a water-sediment mixture. The concentration C_g is expressed as the mass in kg per unit of volume in m^3; the ratio of kg/m^3 can also be expressed as ppm (parts per million). For concentrations lower than 16,000 ppm, it can be assumed that 1 kg/m^3 is equivalent to 1000 ppm.

Both concentrations (by volume and by mass) are related by:

$$C_g = \Delta C \tag{3.21}$$

where:
Δ = relative density (dimensionless)
Δ = 1.65 for quartz particles in water
C = concentration of sediment by volume (percentage or ppm)
C_g = concentration of sediment by mass (percentage or ppm)

CHAPTER 4

Design Criteria for Irrigation Canals

4.1 INTRODUCTION

The objective of an irrigation canal is to meet the varying irrigation requirements during the irrigation season at the individual farms and therefore the canal design should preferably be based on the following criteria:

– *Capacity*: the capacity of an irrigation canal depends on the water delivery method. The delivery methods differ from each other through the scheduling of the water supply in time and place. The World Bank (1986) classifies the following delivery methods:
  - *on demand*: the water delivery reacts instantaneously to the water demand;
  - *continuous*: the irrigation canal supplies a continuous, either constant or varying, flow during the whole irrigation season;
  - *fixed rotation:* the water delivery is scheduled with a constant flow and a regular pattern of rotation;
  - *variable rotation*: the water delivery is programmed either for a fixed supply with variables periods (rotation) or for a variable supply and variable periods;

– *Command*: the requirement that most of the area must be irrigated (commanded) determines the water level in the canals. At any delivery point, the water level in the canal must be above the ground surface. Due to the presence of structures and the need to meet the varying irrigation requirements during the irrigation season the flow in an irrigation canal varies in time and space. The command level is also influenced by the method of flow control. For instance, the water level for downstream control and zero discharge in a canal is horizontal and above the level for full discharge (see Figure 4.1); the downstream control requires canal banks with a *level top* to facilitate the zero flow conditions (Ankum, 1995);

– *Sedimentation and/or erosion*: irrigation canals should be designed based on the criterion that no sedimentation and no erosion occur during a certain period. The design of a stable cross section will be the end result of this criterion;

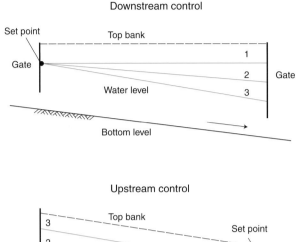

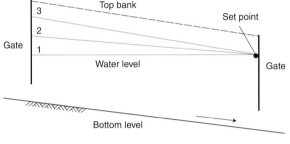

Figure 4.1. Example of a canal with upstream control and a canal with downstream control.

1 $Q = 0$
2 $0 < Q < Q_{design}$
3 $Q = Q_{design}$

– *Cost*: the final result of the design of earthen and lined canals will include the horizontal and vertical canal alignment and the cross sections of the canal. Moreover, the canal design should give the best performance at minimum cost, meaning that the final design should result in a balanced earthwork (including cut and fill) as far as possible.

4.2 FLOW CONTROL SYSTEMS

An irrigation system can be operated in different modes to supply the required water to the irrigation fields. Three parameters can be manipulated to adjust the supply of water; namely the amount, duration and frequency of the supply. The most complex case from a hydraulic and sediment transport perspective is when all three parameters can be changed.

Irrigation canal operations can be broadly divided into:
– *continuous flow*. The situation is a simple one, in which all the canals are opened simultaneously. The available water is proportionally distributed over the whole area;

– *on-off or rotational flow*. Different combinations of rotation are possible. Depending upon the water requirement and available water the secondary canals (laterals) can be grouped in two or more groups and the canals in one group are opened and closed simultaneously;
– *adjustable flow*. The adjustment in flow rate may be necessary in view of a change in demand or change in the available water at the source.

Flow control systems manage the water flow at bifurcations to meet the irrigation service criteria regarding flexibility, reliability, equity and adequacy of delivery (van Lier et al., 1999). Flows can be regulated through water level control, discharge control, and/or volume control. A combination of water level and discharge control is most common in irrigation systems. Flow rates at offtakes are often indirectly controlled through water level control in the conveyance canals. Variation in discharge at the offtake depends on the variations in upstream water level and the sensitivity of the offtakes and water level regulators to those variations.

The classification of the methods of flow control and water division (Ankum, 2004) can be made according to the orientation of the control (upstream, downstream and volume control), the degree of automation (manual, hydraulic or automatic control) and the form of control (local or central control). A short description of upstream and downstream control systems will be given as these systems will be used in the applications with the mathematical model (Chapter 7).

4.2.1 *Upstream control*

Upstream control is most commonly used around the world. Controlled flows are released to a downstream reach from the upstream end according to a pre-arranged schedule. The water level in the reach is maintained by a water level regulator at the downstream end. The offtakes are located upstream of the water level regulator. Since the flow rates are controlled from the upstream end, increments in the offtake flow cannot be diverted unless extra flow is released at the upstream end. This mode of operation forms a negative dynamic storage within the canal reach (Figure 4.2). From the operational point of view, there is no storage available to meet the immediate demand of the offtakes. Moreover, when there is a sudden decrease in the demand and the offtake gates are closed, this stored water has to be released either to the drains or to the downstream reaches that may or may not use the available water. Upstream control is a serial control, is supply oriented, has limited flexibility and requires a well-equipped centralized management to operate the system properly.

Upstream control can be made fully automatic and is an option for semi-demand allocation to the tertiary offtakes. Automation is only suitable when there is sufficient water at the source and when the power supply for the operation of the control structure is reliable.

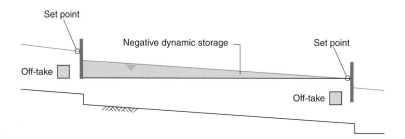

Figure 4.2. Upstream control.

4.2.2 *Downstream control*

Downstream control systems respond to water level changes downstream of a regulator. The structures permit instantaneous response to changes in demand by using the storage in the upstream canal section (positive dynamic storage) in the case of an increase in demand and by storage of water in the case of a decreasing demand. Downstream control systems can be manually operated, but they are easily automated, either hydraulically (Neyrpic) or electrically. AVIS and AVIO are examples of automatic gates that react to the water level at the downstream side (Ankum, 2004).

Both water level and flow are controlled at the upstream end of a canal section (Figure 4.3). Changes are gradually passed on in the upstream direction towards the head works. Most downstream control systems use balanced gates. These control systems are specifically designed to maximize flexibility by minimizing the system response time.

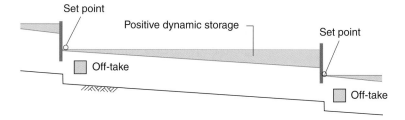

Figure 4.3. Downstream control.

The application of downstream control is usually limited to canals with a bed slope smaller than 0.3‰. Downstream control can be applied both in on-demand and on-request delivery systems. The main design requirement is that each canal section has sufficient capacity to meet the maximum instantaneous demand. Downstream control is also a serial process.

4.3 THE ROLE OF SEDIMENT TRANSPORT IN THE DESIGN OF IRRIGATION CANALS

The design of irrigation canals for sediment-laden water should consider aspects related to the conveyance of irrigation water as well as the transport of sediments. The need to convey different quantities of water to meet the irrigation requirements at the required water level is the main criterion for canal design. Furthermore, the design must be compatible with a particular sediment load in order to avoid silting and/or scouring (Lawrence, 1990 and 1993). The diverted discharge should meet the irrigation requirements and at the same time it should result in the least deposition and/or erosion in the canal network.

Vanoni (1975) stated that the canal design must be based on a comprehensive assessment of the canal operation to determine the future pattern of the water demand. In that way the sediment transport characteristics along a canal network can be established in terms of place and time. Sediment may be deposited during one phase of operation and eroded during another phase, resulting in a balanced or stabilized condition.

According to FAO (1981), the objective of a canal design is to select the proper bottom slope and geometry of the cross section so that during a certain period the sediment flowing in is equal to the sediment flowing out of the canal. Changes that occur in the equilibrium conditions for sediment transport will result in periods of deposition and/or erosion.

Chang (1985) suggested that in view of these sediment problems, the bottom slope and the canal geometry must be interrelated in order to maintain the best possible sediment transport equilibrium. He stated that the sediment problem can be controlled by maintaining the continuity in sediment transport in the irrigation canals during the design stage.

Dahmen (1994) pointed out that an irrigation network should be designed and operated in such a way that:
– the required flow passes at the design water level;
– no erosion of the canal bottom and banks occurs;
– no deposition of sediment in the canal takes place.

The design of a canal that has to convey a certain sediment load requires a set of equations related to the water-sediment flow to provide the unknown variables of bottom slope and cross section (bottom width and water depth). The geometry of an irrigation canal that carries water and sediment will be the end result of a design process in which the flow of water and the transport of sediments interact.

In view of the design, irrigation canals can be divided into three categories (Ranga Raju, 1981) that can be described as follows:
– *canals with a rigid boundary*: the design is based on the determination of the velocity at which any sediment entering into the canal will not settle on the canal perimeter. High velocities are allowed, but they should not damage the lining or create large disturbances in the water surface. A numerical simulation of the most likely changes in the flow conditions during the irrigation season becomes an important tool to ensure that the sediment is not deposited, even when the velocity is low;

– *canals with an erodible boundary and carrying clean water*: the canal design is based on the determination of the maximum velocity for which the bed material in the cross section does not move. The minimum cross section with a maximum velocity that does not result in scouring of the bed should be the end result of the design;

– *canals with an erodible boundary and carrying water with sediment*: the design principle assumes that the canal should transport the water as well as the sediment. The cross section must ensure flow velocities as large as possible to convey the sediment and at the same time not too large to prevent scouring of the bed. It is evident that it is difficult to meet both restrictions simultaneously and for the whole irrigation season. Therefore, the design must look for a stable canal over a longer period and this means that the total sediment inflow during a certain period must equal the total sediment outflow.

A major conclusion to be drawn from the previous section is the need for the design of stable canals during the full operating life of the irrigation network. Sediments may be deposited during one phase of the irrigation season and be eroded during another phase, but for the total operation period, there should be a balance between erosion and deposition in the canal.

Numerous researchers have developed a wide range of canal design methods that are in use around the world. In the field of alluvial canal design the earliest work was of Kennedy (1895). Several others followed with width predictors for alluvial canals. The concepts of minimum stream power (Chang, 1980), minimum energy dissipation (Brebner and Wilson, 1967; Yang et al., 1981) and maximum sediment transport (White et al., 1981) have been proposed to conquer limitations in the regime approach. Lane (1955) presented the tractive force theory developed by many others at the U.S. Bureau of Reclamation. This approach is more suitable for canals that carry very little sediment and the problem is limited to controlling the bed or bank erosion. Depending upon their fundamental solution these methods can be broadly classified under regime, tractive force and rational theories.

Chow (1983), Raudkivi (1990), HR Wallingford (1992), Simons and Sentürk (1992), and others mention four methods for the design of stable canals:
– the regime method;
– the tractive force method;
– the permissible velocity method;
– the rational method.

4.3.1 *Regime method*

The regime design methods present sets of empirical equations based on observations of canals and rivers that have achieved dynamic stability.

An alluvial canal is said to be 'in regime' or in dynamic stability when, over some suitably long period, its depth, width and slope stabilize to average or equilibrium values (Raudkivi, 1990). There may be seasonal deposition or erosion in the canal but the overall canal geometry in one water year remains unchanged. This can only occur when the sediment input to the canal matches the average sediment transporting capacity, sediment deposition during periods of high sediment input being balanced by periods of scour when the sediment input is low.

The regime method considers the three main features of a canal, namely the perimeter of the open canal, the amount of water and the sediment flowing in it, as a whole. It attempts to derive the most crucial characteristics of a stable (non-silting and non-scouring) canal primarily on the basis of field and laboratory studies that investigated the interaction of the above-mentioned factors (Naimed, 1990). The regime equations are based on past and present observations and experiences and they present a long-term average rather than an instantaneously variable state. Therefore, the regime method tries to express the natural tendency of canals that convey sediment within alluvial boundaries, in order to seek a dynamic stability.

Canals are described as in regime if they do not change over a period of one or more typical water years. Within this period, scour and deposition are allowed to occur as long as they do not interfere with canal operations. Since the observed canals withdraw varying amounts of water and sediment from different rivers and since sediment excluders may be in operation at some of the head works, the regime theory can only provide some approximate average design values. Nevertheless, the adequate experience obtained from the design and operation of these canals give some proficient guidance for the design of stable channels with erodible banks and sediment transport. However, the applicability of this method can be challenged in the case of highly time-dependent operational regimes as practised in many irrigation systems at present (Bruk, 1986).

Another limitation of the regime theory is that it assumes that the discharge is the only factor determining the wetted perimeter while the fact is that canals with less stable banks tend to be wider than those with strong banks for the same discharge.

Regime methods were developed at the end of the 19th century to aid the design of major irrigation systems on the Indian sub-continent. The development started with Kennedy (1895) followed by Lindley (1919), Lacey (1930), Blench (1957, 1970), and Simons and Albertson (1963). A critical analysis of these methods is given by Stevens and Nordin (1987) and they are well explained in the books by Shen (1976), Chang (1988), Raudkivi (1990) and Simons and Senturk (1992).

One of the most popular regime methods is the set of equations of Lacey (1930). The equations were based on data from three canal systems on the Indian sub-continent. They specify the cross section and slope of regime canals based on the incoming discharge and a representative bed material size (Chitale, 1994). These equations, with minor changes to coefficients and some redefinition of the silt factor, are still widely used.

Some of the equations given by Lacey are (Ackers, 1992):

$$f = \sqrt{2520d} \tag{4.1}$$

$$P = 4.836Q^{0.5} \tag{4.2}$$

$$v = 0.6459(fR)^{0.5} \tag{4.3}$$

$$S_{\mathrm{o}} = 0.000315\frac{f^{5/3}}{Q^{1/6}} \tag{4.4}$$

where:
f = Lacey's silt factor for a sediment size d
d = sediment size (m)
P = wetted perimeter (m)
Q = discharge (m^3/s)
R = hydraulic radius (m)
v = mean velocity (m/s)
S_{o} = bottom slope

According to Lacey (1958), these equations are applicable within the following range of flow and sediment characteristics:
- Discharge in the range of 0.15–150 m^3/s
- Bed material is non-cohesive
- Bed form characterised by ripples
- Bed material size in the range of 0.15–0.40 mm
- Bed load is relatively small

The following steps are recommended when using the Lacey equations for the design of earthen canals for a given diameter d and discharge Q:
- Determine the factor f for the given diameter d;
- Find the bottom slope S_{o} and the perimeter P;
- By trial and error determine the velocity v and area $A = Q/v$ and the hydraulic radius R;
- Determine the cross section both by using the perimeter P and the hydraulic radius R and by assuming that the cross section is trapezoidal with a side slope $m = 2$ (1 V : 2 H).

Table 4.1 gives an example of the main dimensions of an earthen canal designed according to the Lacey method for a canal that conveys a discharge in the range from 1–8 m^3/s and that transports sediment with a diameter $d = 0.001$ and 0.0005 m, respectively. The f values for these diameters are 1.58745 and 1.1225, respectively. The side slope m is 2.

As an extra illustration, the table also gives the bottom width B and the water depth y as well as the Manning's roughness coefficient that has been calculated from the equation

$$v = \frac{1}{n}R^{2/3}S_{\mathrm{o}}^{1/2}$$

Table 4.1. Example of the design of an earthen canal according to the Lacey method.

Lacey regime theory in SI-units

Q	d in m	m	f	S_0	P	v	A	R	B-bottom	y	n-Manning
1	0.001	2	1.58745	0.000680	4.836	0.515	1.940	0.401	2.316	0.563	0.028
2	0.001	2	1.58745	0.000606	6.839	0.579	3.457	0.505	3.863	0.666	0.027
3	0.001	2	1.58745	0.000567	8.376	0.619	4.846	0.579	5.065	0.740	0.027
4	0.001	2	1.58745	0.000540	9.672	0.649	6.159	0.637	6.091	0.801	0.026
5	0.001	2	1.58745	0.000520	10.814	0.674	7.418	0.686	7.004	0.852	0.026
6	0.001	2	1.58745	0.000505	11.846	0.695	8.635	0.729	7.835	0.897	0.026
7	0.001	2	1.58745	0.000492	12.795	0.713	9.819	0.767	8.604	0.937	0.026
8	0.001	2	1.58745	0.000481	13.678	0.729	10.975	0.802	9.324	0.974	0.026
1	0.0005	2	1.12250	0.000382	4.836	0.459	2.178	0.450	1.693	0.703	0.025
2	0.0005	2	1.12250	0.000340	6.839	0.515	3.880	0.567	3.275	0.797	0.025
3	0.0005	2	1.12250	0.000318	8.376	0.551	5.440	0.649	4.459	0.876	0.024
4	0.0005	2	1.12250	0.000303	9.672	0.579	6.914	0.715	5.463	0.941	0.024
5	0.0005	2	1.12250	0.000292	10.814	0.600	8.327	0.770	6.353	0.997	0.024
6	0.0005	2	1.12250	0.000283	11.846	0.619	9.693	0.818	7.163	1.047	0.024
7	0.0005	2	1.12250	0.000276	12.795	0.635	11.021	0.861	7.913	1.092	0.024
8	0.0005	2	1.12250	0.000270	13.678	0.649	12.319	0.901	8.614	1.132	0.024

For increasing discharge Q, the bed width B, the water depth y and the B/y-ratio will increase. The bottom slope S_0 will decrease with an increase of the discharge Q. The regime theory results in relatively wide and shallow cross sections.

Simons and Albertson (1963) developed a set of regime equations based on a large data set collected from canal systems in India and North America. Five types of canals were identified from the collected data set and for each category the coefficients are given for the design of a canal, as shown in Table 4.2. Simons and Albertson's equations have a wide range of applicability and the data set used is related to a sediment load of less than 500 ppm (Raudkivi, 1990). The set of regime equations of Simons and Albertson (1963) is:

$$P = K_1\sqrt{Q} \tag{4.5}$$

$$B = 0.9P = 0.92B_s - 0.61 \tag{4.6}$$

$$R = K_2 Q^{0.36} \tag{4.7}$$

$$h = 1.21R \quad \text{for } R < 2.1\,\text{m} \tag{4.8}$$

$$h = 0.61 + 0.93R \quad \text{for } R > 2.1\,\text{m} \tag{4.9}$$

$$v = K_3(R^2 S)^n \tag{4.10}$$

$$\frac{C^2}{g} = \frac{v^2}{gDS} = K_4 \left(\frac{vB}{v} \right)^{0.37}$$

(4.11)

$$A = K_5 Q^{0.87}$$

(4.12)

where:
A = cross sectional area of the flow (m^2)
B = mean canal width (m) = A/h
B_s = water surface width (m)
C = Chézy coefficient (m$^{1/2}$/s)
h = mean flow depth (m)
K_i = coefficient (i = 1 to 5) for different canal types
n = exponent
P = wetted perimeter (m)
Q = flow rate (m^3/s)
R = hydraulic mean radius (m)
S = bed slope (m/m)
v = flow velocity (m/s)
v = kinematic viscosity (m^2/s)

Table 4.2. Value of coefficients in Simons and Albertson's equations for different canal types (Simons and Albertson, 1963).

Canal type	Coefficient					
	K_1	K_2	K_3	K_4	K_5	n
1. Sand bed and banks	6.34	0.4–0.6	9.33	0.33	2.6	0.33
2. Sand bed and cohesive banks	4.71	0.48	10.80	0.53	2.25	0.33
3. Cohesive bed and banks	4.0–4.7	0.41–0.56	–	0.88	2.25	–
4. Coarse non-cohesive material	3.2–3.5	0.25	4.80	–	0.94	0.29
5. Sand bed, cohesive bank and heavy sediment load*	3.08	0.37	9.70	–	–	0.29

*Sediment (mainly wash load) 2 000 to 3 000 ppm.

Regime theories are available in many areas where irrigation is practiced, but the fact that these methods are not being transformed to other places is an indication that not all the physical parameters defining the problems are correlated by the regime methods (Raudkivi, 1990 and Bakker et al., 1989).

4.3.2 Tractive force method

The tractive force methods are used for the relationship between the boundary shear stress and sediment transport. They use the concept of static stability of a canal in which there is no movement of material (either

on the bed or side slopes). For a given design discharge, the canal dimensions and bed slope are determined considering the flow velocity not exceeding a permissible velocity or boundary shear stress. These values are related to the critical values for the bed and bank material. Under the tractive force theory two methods are in use:
– maximum permissible velocity;
– critical shear stress.

A summary of the available equations for the prediction of velocity is given by Raudkivi (1990) and Simons and Senturk (1992).

The minimum velocity in an irrigation canal should not induce sedimentation and at the same time the velocity should limit unwanted aquatic weed growth and reduce health risks (for example, schistosomiasis). The minimum velocity is a function of the shape of the canal. For large canals, a minimum velocity of 0.30 m/s is recommended (Dahmen, 1999). Smaller velocities result in uneconomic large cross sections. A velocity of 0.10 to 0.15 m/s is recommended for minor canals (tertiary and small secondary canals); smaller velocities result in uneconomic wide sections.

The maximum permissible velocity should not cause erosion of the bottom and side slopes and a critical shear stress criterion is applied, which is extremely dependent on the fact that:
• the resistance to erosion increases when smaller particles are washed out;
• an aged canal has more resistance to erosion than a newly constructed one;
• colloidal matter in the water will increase the cohesion of the particles that form the boundaries, resulting in a larger resistance;
• a higher ground water table than the canal water level will decrease the resistance; a lower ground water level will increase the resistance.

The tractive force depends on the shear stress at the bottom, which can be expressed as (Dahmen, 1994):

$$\tau = c\rho g y S_o \tag{4.13}$$

with:
τ = tractive force per unit wetted area (N/m^2)
c = correction factor; the correction factor c depends on the B/y ratio:
$c = 0.77e^{0.065(B/y)}$ for $1 < B/y < 4$
$c = 1$ for $B/y \geq 4$
$\rho = \rho_w$ = density of water (1000 kg/m^3)
S_o = bed slope (m/m)
y = water depth (m)
g = acceleration due to gravity (m/s^2)

The tractive force concept primarily originates from work carried out by the U.S. Bureau of Reclamation (USBR) under the direction of Lane (Raudkivi, 1990 and HR Wallingford, 1992). This method considers the balance of forces acting on sediment grains and is only used to evaluate the erosion limits. Since the method assumes no suspended or bed material transport, it is only relevant for canals with coarse bed material and zero or very small bed material input (HR Wallingford, 1992). Breusers (1993) also supported the statement that the tractive force method is only suitable when the water transports very little or no sediment.

The distribution of the shear stress in narrow and trapezoidal canals is not uniform and the maximum shear stress is smaller than the one predicted by Equation 4.13. Lane (1953) determined experimentally that the adjustment factor for both the bed and side slopes largely depends on the width to depth ratio B/y and side slope m (Figure 4.4). Based on extensive work and field data of the U.S. Bureau of Reclamation, Lane (1955) established a critical tractive force diagram that relates the critical force value with the mean diameter of the bed material for canals carrying water with different amounts of sediment.

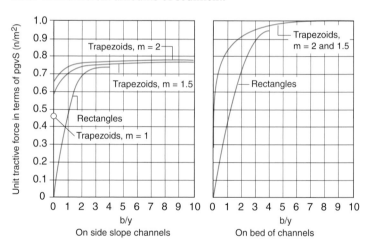

Figure 4.4. Maximum shear stress in a canal (Lane, 1953) (adapted from Chow (1983)).

These diagrams together with the uniform flow equation (Manning or de Chézy) can be used to design a stable canal. Normally the grain size (d_{50}) of the bed material is known, so the design process involves the assumption of any three of the side slope, bed slope, bed width, water depth and B-y ratio to find the remaining two. The merits and limitations of this method have been discussed by Simons and Albertson (1963).

The tractive force method was developed for the threshold condition of sediment transport, which occurs for the critical boundary shear stress along the canal's perimeter. Field studies on very coarse material showed that the 'critical shear stress', above which motion would start, is approximately $0.94 * D_{75}$ (N/m^2). For the canal design a boundary shear stress of 0.80% of the critical shear stress is recommended: $\tau_{\text{design}} = 0.75 * D_{75}$ (D_{75} in mm). The allowable shear stress is a function of the mean diameter and the sediment concentration of the water (see Table 4.3).

Table 4.3. Recommended critical shear stress (N/m^2) for fine, non-cohesive sediment (Dahmen, 1994).

D_{50} (mm)	Clear water	Light load	Heavy load
0.1	1.20	2.40	3.60
0.2	1.25	2.49	3.74
0.5	1.44	2.64	3.98

Dahmen (1994) suggested as *a rule of thumb* for many irrigation engineers, that the maximum boundary shear stress in 'normal soil', for a 'normal canal' and under 'normal conditions' can be between 3 and 5 N/m^2. When the shear stress is too high, the most sensitive factor to reduce this factor is a gentler bottom slope. Table 4.3 shows some recommended values for the critical shear stress (N/m^2) along the boundary for fine, non-cohesive sediment.

For cohesive soils no definite criteria exist, but based on field data, the USBR gives the following values (see Table 4.4).

Table 4.4. Limiting boundary shear stress (N/m^2) in cohesive material.

Description	Compaction			
	Loose	Fair	Well	Very well
Void ratio	2.0–1.2	1.2–0.6	0.6–0.3	0.3–0.2
Cohesive bed material	Limiting boundary shear stress (N/m^2)			
Sandy clays	1.91	7.52	15.7	30.2
Heavy clay soils	1.48	6.75	14.6	27.0
Clays	1.15	5.94	13.5	25.4
Lean clayey soils	0.957	4.60	10.2	16.9

The shear stress on the side slopes is only considered when the bed and sides are covered with coarse, non-cohesive material. For this situation the shear stress together with the angle of repose has to be taken into account. For fine, non-cohesive material and for cohesive material the component of the weight is small and can be ignored.

The values of the critical boundary shear stress are established for the design of straight canals and they should be reduced for sinuous canals; Table 4.5 presents some reduction values (in %) for non-straight canals.

Table 4.5. Reduction of the limiting boundary shear
stresses in non-straight canals.

Slightly sinuous	10%
Moderately sinuous	25%
Very sinuous	40%

As discussed in this chapter, the design of irrigation canals needs at least four equations to establish the main dimensions of the canal. For earthen canals the side slope will be based on the (estimated) canal depth and soil properties and therefore, three equations are required to determine the remaining variables. Hence, the application of the tractive force method requires two other equations: for example one equation to compute the discharge (Manning, Strickler or de Chézy) and another for the relationship between the bottom width and the water depth (B/y).

The hydraulically optimal cross-section has a minimized perimeter P, which results in a maximum flow velocity at minimum cost. However, the optimum hydraulic section is hardly ever applied; this is because it will not be stable due to relatively deep excavations and also any change in discharge severely affects the water depth and the velocity. A deep section is nevertheless applied wherever possible, because the expropriation costs will be less, the velocity is higher in a deep rather than in a shallow canal and the sediment transport capacity is larger in deeper canals (the transport capacity is linear with the bottom width, but exponential with the water depth). To limit the excavation and expropriation cost, canal side slopes are designed to be as steep as possible. Soil material, canal depth and the danger of seepage determine the maximum slope of a stable side slope. The side slope has to be stable under *normal conditions*, also against erosion. For deep excavations, an extra berm can be included to improve the slope stability. Table 4.6 presents some values for side slopes in irrigation canals.

Table 4.6. Side slopes in canals.

Material	Side slope: 1 V : m H
Rock	0.0
Stiff clay	0.5
Cohesive medium soils	1.0–1.5
Sand	2.0
Fine, porous clay, soft peat	3.0

The top width of canal embankments depends on the soil type, the side slope and special requirements in view of maintenance and operation. Water levels may rise above the design water level due to deterioration of the canal embankment, a sudden closure of a gate or unwanted drainage inflow. A freeboard, being the distance between the design water level and the canal bank, is provided to safeguard the canal against *overflow-ing* during unexpected water level fluctuations and wave actions. USBR recommends a freeboard $F = Cy$, where the coefficient C varies between 0.5 to 0.6 with a minimum of 0.15 or 0.20 m.

4.3.3 *Permissible velocity method*

Depending on the fact whether a canal will be erodible or non-erodible, the permissible velocity concept can be used as a design criterion for stable canals.

- The minimum permissible velocity is defined as that velocity that will not result in sedimentation or induce the growth of aquatic weeds. This velocity depends on the sediment transport capacity of the canal.
- The maximum permissible velocity is that velocity that will not cause erosion. This velocity is difficult to ascertain and is very variable; it can only be estimated with experience and sound judgment (Chow, 1983).

Table 4.7 presents the maximum, permissible velocities depending on the bed material (Simons and Sentürk, 1992). The USBR has derived from these velocities the corresponding tractive force values (Chow, 1983).

Table 4.7. Maximum permissible velocities and the corresponding tractive force values.

Material	Manning n	Clear water		Silt-loaded water	
		v (m/s)	τ (N/m^2)	v (m/s)	τ (N/m^2)
Fine sand, colloidal	0.02	0.46	1.30	0.76	3.61
Sandy loam, non-colloidal	0.02	0.53	1.78	0.76	3.61
Silt loam, non-colloidal	0.02	0.61	2.31	0.91	5.29
Alluvial silts, non-colloidal	0.02	0.61	2.31	1.07	7.22
Ordinary firm load	0.02	0.76	3.61	1.07	7.22
Volcanic ash	0.02	0.76	3.61	1.07	7.22
Stiff clay, very colloidal	0.025	1.14	12.51	1.52	22.13
Alluvial silts, colloidal	0.025	1.14	12.51	1.52	22.13

4.3.4 *Rational method*

The design of irrigation canals typically tries to find the four unknown dimensions, which include the bottom slope S_o, bottom width B, water depth y and side slope m. The side slope m should be based on the soil properties and the (estimated) canal depth y. Hence, three equations are required to determine the remaining variables. They might include an alluvial friction predictor (de Chézy, Manning, and Strickler), a sediment transport predictor and a minimum stream power or maximum sediment transport efficiency. Sometimes a regime relationship is used to provide a relationship for bottom width and water depth (HR Wallingford, 1992).

The rational method includes, amongst others, the method of White, Bettes and Paris (1982) and Chang (1985). The rational method is useful for the design of stable canals with very specific flow conditions. For canals with large variations in discharge and sediment load the method is inadequate to describe accurately the sediment transport process and the conveyance of the sediment load through the whole canal network.

Transport of suspended load

To describe the conveyance of suspended sediment through the whole network it is practical to assume that the fine particles in suspension have an almost constant distribution over the vertical. Normally the sediment particles in suspension fall in the range 'very fine' and their size is less than 50 to $70 * 10^{-3}$ mm. De Vos (1925) stated that the relative transport capacity (T/Q) is proportional to the average energy dissipation per unit of water volume.

$$\frac{T}{Q} \propto \rho_{\mathrm{w}} g v_{\mathrm{av}} S_{\mathrm{o}} \tag{4.14}$$

where:
T/Q = relative transport capacity
T = sediment transport load (m³/s)
Q = discharge (m³/s)
ρ_{w} = density of water (kg/m³)
g = acceleration due to gravity (m/s²)
v_{av} = average velocity (m/s)
S_{o} = bottom slope (m/m)

From energy considerations it follows that the sediment particles will be transported in any concentration by the flowing water when the fall velocity w is smaller than a certain threshold.

$$w \leq \left(\frac{\rho_{\mathrm{w}}}{\rho_{\mathrm{s}} - \rho_{\mathrm{w}}} \right) v_{\mathrm{av}} S_{\mathrm{o}} \tag{4.15}$$

where:
w = fall velocity (m/s)
ρ_{s} = density of the sediment particles (kg/m³)
ρ_{w} = density of water (kg/m³)
S_{o} = bottom slope (m/m)

To convey sediment in suspension the hydraulic characteristics of a canal network should be such that $\rho_{\mathrm{w}} g v_{\mathrm{av}} S_{\mathrm{o}}$ or $v_{\mathrm{av}} S_{\mathrm{o}}$ either remains constant or does not decrease in a downstream direction. From $v_{\mathrm{av}} = C(y S_{\mathrm{o}})^{0.5}$ in which C is a general smoothness factor, it follows that $y S_{\mathrm{o}}^3$ or $y^{1/3} S_{\mathrm{o}}$ should be constant or non-decreasing in order to convey the suspended load in wide canals.

Transport of bed load

For canal networks that transport some material as bed load, it is possible to establish a comparable criterion as for the transport of suspended load. Irrigation canals that carry substantial sediment loads cannot be treated either by the rational method or in this way.

Bed load transport will mainly occur on or above the canal bottom; almost no transport will occur on the banks. The bed load might be some sediment with a diameter larger than 50 to 70 * 10^{-3} mm. The amount of bed load transport mainly depends on the shear velocity $v_* = (gyS_o)^{0.5}$. The continuous transport of bed load material can be described by one of the various sediment transport formulae, for example Engelund Hansen, Ackers-White, Brownlie, etc.

The relative transport capacity is:

$$\frac{T}{Q} \propto \frac{b * y^3 * S_o^3}{b * y * y^x * S_o^z} \tag{4.16}$$

$$\frac{T}{Q} \propto y^{2-x} S_o^{3-z} \tag{4.17}$$

where:
T/Q = relative transport capacity
T = sediment transport load (m^3/s)
Q = discharge (m^3/s)
b = bottom width (m)
y = water depth (m)
S_o = bottom slope (m/m)
x, z = exponents depending on the choice of the water flow equations, for instance Manning, Strickler, de Chézy or Lacey. For example, the exponents for the regime theory are $x = 0.75$ and $z = 0.5$; for the Strickler equation they are $x = 3/2$ and $z = 0.5$.

To prevent sedimentation the relative transport capacity $T/Q = y^{2-x} S_o^{3-z}$ should either be constant or not decrease in the downstream direction of the irrigation network. For $x = 0.75$ and $z = 0.5$ the relationship T/Q is proportional with $y^{0.5}S_o$. Also with other flow equations (Manning, Strickler or de Chézy) and sediment transport predictors (for instance, Engelund-Hansen or Einstein-Brown), the relative sediment transport capacity proves to be almost proportional with $y^{0.5}S_o$.

To prevent erosion, the boundary shear stress or the shear velocity should not increase, but should remain constant or decrease in a downstream direction.

The shear velocity follows from:

$$v_* = \sqrt{gyS_o} \tag{4.18}$$

In order to have no erosion in the canal network the shear stress should be constant or non-decreasing, meaning that yS_o should be constant or non-decreasing.

It might be expected that when a flow only transports a small amount of bed load, the criterion for a continuous conveyance of bed load should be somewhere in between the two values. The criterion to convey any non-suspended (bed) material depends strongly on the water and sediment equations. Therefore, the criterion is that the relative transport capacity for bed load (T/Q) should be non-decreasing, or in the case of potential erosion, should remain constant; the numeric approximation for this bed load transport is that $y^{1/2}S_o$ is constant or non-decreasing.

Summarizing the rational method criteria
The design criterion for the conveyance of sediment through a canal system can be based on energy dissipation considerations; the relative sediment transport capacity follows from these considerations and is given by de Vos (1926) and Vlugter (1962).

Based on their works, Dahmen (1994) states that:
- for the conveyance of sediment in suspension, the hydraulic characteristics of the canal system should be such that:

$$\rho g v S_o = \text{constant or non-decreasing in downstream direction} \quad (4.19)$$

- for the conveyance of non-suspended sediments, the hydraulic characteristics of the canal system should be such that:

$$y^{1/2} S_o = \text{constant or non-decreasing in downstream direction} \quad (4.20)$$

4.3.5 *Final comments*

The preceding sections clearly show that the design of stable irrigation canals should be based on irrigation considerations, including engineering, agricultural, management and economic aspects. An optimal canal design is difficult to achieve and the final canal design will have to balance all these criteria to find the best solution for the explicit conditions of a specific irrigation system.

The existing design methods are based on the interrelation of equations for specific water flow and sediment transport conditions in an effort to design stable canals. However, the input variables will vary widely during the irrigation season and, moreover, during the lifetime of the irrigation network. Most of the time, non-equilibrium conditions prevail in irrigation networks and therefore, the initial assumptions for a stable canal design are no longer valid. Also, lined canals experience sedimentation problems due to variations in either flow conditions or incoming sediment load that produce non-equilibrium conditions for the transport of the sediment. Therefore, sediment problems in irrigation canals should be analysed in a more integrated context, including all the alternative operation scenarios for water flow and sediment transport in time and space.

The canal design methods as discussed above do not directly use available information on sediment characteristics, such as sediment size and concentration. Nowadays the design of irrigation canals is the product of complex and difficult to determine parameters such as water flow, required water levels, sediment load, structure control and operation and management strategies. No design packages are available that deal with all these parameters at the same time. When analyzing one parameter, others are either ignored or assumed to be constant. Therefore, it is obvious that a mathematical model for the specific conditions of irrigation canals will be an important tool for designers and managers of irrigation systems.

4.4 APPURTENANT CANAL STRUCTURES

4.4.1 *Irrigation structures*

The conveyance of irrigation water from the head works to the farmer's field can be realized by an open irrigation network that consists of canals with a large number of appurtenant structures; the latter can be divided into a conveyance and an operational part. The conveyance part includes the canals and structures that convey, regulate, measure and divide the water flow. The operational part consists of the structures that divide and control the flow in terms of water level and/or discharge; they may be without any movable part (fixed) or they have movable parts such as gates for the control.

The main objectives of an irrigation network are to deliver irrigation water:
- in the right quantity (in size, frequency and duration);
- at the required level (head);
- at the right place;
- at the right moment;
- equitably (fairly and objectively);
- in a reliable and assured way.

To meet these objectives the structures will have to regulate, control, measure or distribute the water flow (Leliavsky, 1983). The design of the conveyance part is normally based on hydraulic and structural requirements, local conditions, available technology and cost. The operational part includes additional structures that support the operation and management for the water delivery mode, acceptance by the water users, ease in operation, transparency, etc. The sediment transport aspect is normally not considered in the selection and design of canal structures. An irrigation system can comprise the following structures:
- head works at the head of a primary or main canal: a weir or barrage, free intake, reservoir, pumping station;
- conveyance structures: aqueducts, chutes, closed canals, culverts, drop structures, flumes, inverted siphons, siphons, tunnels;

- regulating structures: control structures and regulators maintain the water level for different discharges: checks, regulating gates, control notches and weirs, division works, lateral intake structures, proportional distributors;
- measuring structures for an accurate measurement of the irrigation discharges: a broad-crested weir, Parshall flume, Cipoletti weir, Crump-de Gruyter orifice, baffle distributor, gated pipe offtake, constant head orifice;
- protective structures: culverts, inverted siphons, overchutes, spillways, waste ways, side drainage, sediment traps and settling basins;
- division structures at the head of (sub-)secondary canals;
- offtake structures or turnouts at the outlets to tertiary or end units: tertiary offtakes, turn-outs, pipe offtakes;
- auxiliary structures: appurtenant canal structures, dikes, roads, operation facilities;
- miscellaneous structures: culverts, farm and other bridges, drainage inlets, fish-control structures.

From an operational point of view four major types of structures will be classified:

- fixed (weirs and orifices);
- on-off (shutter gates);
- adjustable: stepwise (stop logs, modular distributors) or gradually adaptable (undershot gates, movable weirs);
- automatic (upstream and downstream water level control structures).

4.4.2 *Main hydraulic principles for irrigation structures*

The type of flow over an irrigation structure can be either overflow (weirs, drops, flumes) or undershot (orifices, gates).

Flow through irrigation structures can be a sub-critical flow (aqueduct, flume, and culvert) or a critical flow (measuring structure, flow control and division structure, drop). The design of a structure for sub-critical flow aims at a minimal head loss with smooth transitions to minimize the entry and exit losses. The flow through a structure with critical flow is either critical or supercritical.

Many situations in hydraulics can be solved by applying two out of the three main principles:

1. conservation of matter (continuity)
2. conservation of energy (all energy losses are known)
3. momentum principle (all external forces are known)

The continuity principle for a structure without any extra outflow or inflow inside the structure results in the observation that the upstream incoming flow is equal to the downstream outgoing flow:

$$Q = \bar{v}_1 A_1 = \bar{v}_2 A_2 \tag{4.21}$$

The energy principle holds true as long as proper allowance is made for energy losses (dissipation of energy). In a steady flow with straight and parallel streamlines, the centripetal acceleration is negligible and the sum of the potential and pressure energy at any point gives the total energy per unit of weight:

$$E = \alpha \frac{\bar{v}_1^2}{2g} + \left(\frac{p}{\rho g} + z\right)_1 = \alpha \frac{\bar{v}_2^2}{2g} + \left(\frac{p}{\rho g} + z\right)_2 \quad \text{(Bernoulli equation)}$$

(4.22)

In a real fluid the velocity head is multiplied with the Coriolis coefficient α, and a loss of energy (due to friction and/or local losses) has to be added:

$$\frac{\alpha \bar{v}_1^2}{2g} + \frac{p_1}{\rho g} + z_1 = \frac{\alpha \bar{v}_2^2}{2g} + \frac{p_2}{\rho g} + z_2 + h_f$$

(4.23)

The momentum principle is based upon the momentum passing a cross section per unit of time: $mv = \rho \beta Q v$ where β is the Boussinesq coefficient. The change of momentum per unit of time is equal to the resultant of all external forces acting on a body of flowing water during time dt. Between two cross sections the change of momentum is $\Sigma P = \rho Q(v_2 - v_1)$. For any cross section and assuming hydrostatic pressure the force $P = \rho g A y_1^+$.

$$\frac{P}{\rho g} = \left(A_1 y_1^+ \frac{\beta_1 Q^2}{A^2 g}\right) - \left(A_2 y_2^+ \frac{\beta_2 Q^2}{A^2 g}\right)$$

(4.24)

4.4.3 *Hydraulics for some irrigation structures*

For the conveyance of irrigation water from its source to the farmers field an irrigation network is needed that consists of a large number of appurtenant structures that can be divided into a conveyance and an operational part. The conveyance part includes the canals and fixed structures such as aqueducts, siphons, bridges, culverts, drops and cascades, flow measuring structures, etc. The operational part includes the structures that divide and control the flow in terms of water level or discharge; they may be fixed without any movable part for the control or movable such as gates.

The design of the conveyance part is normally based on hydraulic and structural requirements, local conditions, available technology and cost. The design of the operational part is based upon additional factors that influence the operation and management e.g. water delivery mode, available manpower, acceptance by the users, ease in operation, transparency, etc. In some designs and layouts different structural arrangements are used within the same scheme.

From an operational point of view four major types of structures exist:
- fixed (weirs and orifices);
- on-off (shutter gates);
- adjustable: stepwise (stop logs, modular distributors) or gradually (undershot gates, movable weirs);
- automatic (automatic upstream and downstream water level control structures).

From the sediment transport aspect, irrigation structures can be divided into two categories:
- *structures with sub-critical flow*: examples are aqueducts, flumed canal sections, super-passages, culverts, etc.; the design of structures with sub-critical flow aims at a minimal head loss with smooth transitions to minimize the entry and exit losses.
- *structures with critical flow*: the flow through these structures is either critical or supercritical; examples are measuring structures, flow control and division structures, drops, etc.; structures with critical flow have a unique depth discharge relation when the flow at the control section is critical or modular and the downstream water level will not influence the upstream water level. If the downstream water influences the upstream water level due to the sediment deposition, poor canal maintenance or ponding up of the downstream canal reach, it turns into a drowned or submerged flow.

Structures may have a free overflow or an undershot flow. Some free overflow structures may be used to convey or regulate the water level or measure the discharge. Structures with critical flow present a unique head discharge relationship when the flow at the control section is critical or modular; this is the case when the downstream water level does not influence the upstream water level. The flow turns into a drowned or submerged flow once the downstream water influences the upstream water level due to the sediment deposition, poor canal maintenance or ponding up of the downstream canal. Some important overflow structures are a broad crested weir, sharp crested and short crested weirs and flumes. Hydraulic references give more detailed information on these structures.

The head-discharge equation for a broad-crested weir (Figure 4.5) with a rectangular cross-section and free flow conditions ($h_2 < 0.7h_1$) is (Bos, 1989):

$$Q = C_d B \sqrt{\frac{2}{3}g} \, \frac{2}{3} H_1^{3/2} \tag{4.25}$$

Sometimes it is difficult to estimate the total head (H) and the water depth h_1 is used:

$$Q = C_d C_v B \sqrt{\frac{2}{3}g} \, \frac{2}{3} h_1^{3/2} \tag{4.26}$$

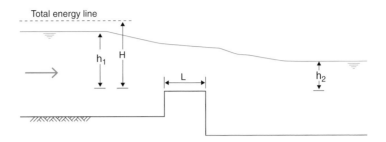

Figure 4.5. Broad crested weir
(Paudel, 2009).

The discharge and velocity coefficient for a round nosed, rectangular broad crested weir are (Bos, 1989):

$$C_d = 0.93 + 0.1\frac{H_1}{L} \quad \text{with } 0.08 < H_1/L < 0.7 \tag{4.27}$$

$$C_v = \left(\frac{H_1}{h_1}\right)^u \tag{4.28}$$

where:
$u = 1.5$ for a rectangular cross section.

For most earthen irrigation canals the Froude number is less than 0.5 and the velocity coefficient C_v is less than 1.05, and the discharge equation for a broad crested weir becomes:

$$Q = CBh_1^{2/3} \tag{4.29}$$

where:
Q = discharge in m^3/s
C = coefficient = $1.705C_dC_v$ and ranges from 1.585 to 1.79
B = width of the weir (m)
H_1 = upstream energy head above the crest (m)
h_1 = upstream water depth above the crest (m)
h_2 = downstream water depth above the crest (m)

A simple sharp-crested weir or notch (see Figure 4.6) has a horizontal edge over the full canal width and a two-dimensional flow without side contraction (suppressed weir). The flow is 'modular' or 'free' when the downstream water level is below the crest. Sharp-crested weirs are not very common in irrigation canals.

$$Q = C_d B\frac{2}{3}\sqrt{2g}h^{1.5} \tag{4.30}$$

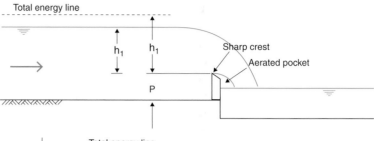

Figure 4.6. Sharp crested weir
(Paudel, 2010).

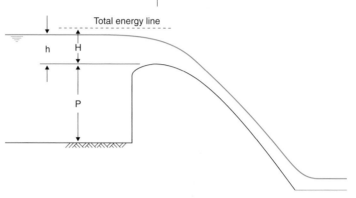

Figure 4.7. Short-crested weir
(Paudel, 2010).

where:

Q = discharge (m^3/s)

C_d = discharge coefficient and is a function of h/P

B = crest width (m)

h = water depth over the crest (m)

H = upstream energy head above the crest (m)

P = height of crest above the upstream bed (m)

Any side contraction (contracted weir) is corrected by a contraction coefficient C_c. For small values of v_0, C_d becomes equal to C_c. An empirical relation for C_d was derived by Rehbock (Henderson, 1966):

$$C_d = 0.611 + 0.08\frac{h}{P} \tag{4.31}$$

The theory for sharp-crested weirs forms the basis for the design of short-crested weirs (spillways) (Bos, 1989). The curvature of a short-crested weir has significant influence on the head discharge relationship (Bos, 1989). The head discharge equation for a short-crested weir (Figure 4.7) with a rectangular control is similar to a sharp-crested weir. Different spillways are in use, such as a WES spillway, ogee spillway and cylindrical-crested weirs. In irrigation schemes they are mostly used as escapes, drops or check structures (Bureau of Reclamation, 1987). For a rectangular control section the head discharge equation is (Bos, 1989):

$$Q = CB_c\frac{2}{3}\sqrt{2g}H^{3/2} \tag{4.32}$$

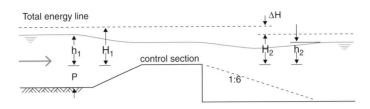

Figure 4.8. Typical rectangular
flume (Wahl, 2005).

For $H = h_{\mathrm{d}}$ (design head), the equation reduces to:

$$Q = 2.20 B_{\mathrm{c}} h^{3/2} \tag{4.33}$$

where:
$Q = $ discharge (m^3/s)
$C = 0.75$ (for the design head above the crest)
$B_{\mathrm{c}} = $ crest width (m)
$H = $ total head above the crest (m)

Flumes consist of an approach canal, a converging transition, usually a throat and a diverging transition (see Figure 4.8). A flume is formed either by narrowing the width or by narrowing the width and raising the bed; in both cases it is called a flume. When only the bed is raised then it is called a weir (Clemmens et al., 2001). Examples of flumes are Parshall, Venturi, long-throated (RBC flume) and cut-throat flumes.

The discharge equation is identical to that of a broad-crested weir (equation).

$$Q = C_{\mathrm{d}}(B_{\mathrm{s}} h_{\mathrm{c}} + m h_{\mathrm{c}}^2)\sqrt{2g(H_1 - h_{\mathrm{c}})} \tag{4.34}$$

where:
$Q = $ discharge (m^3/s)
$C_{\mathrm{d}} = $ discharge coefficient
$B_{\mathrm{s}} = $ bottom width of the sill at the throat (m)
$H_1 = $ upstream energy head (m)
$h_{\mathrm{c}} = $ critical depth in the throat (m)
$m = $ side slope (1 vertical : m horizontal)

The discharge coefficient C_{d} depends upon the size and shape of the flume and for each flume type head discharge relationships are available (Bos, 1989; Bureau of Reclamation, 2001). Flumes may have two

Downstream water level below crest

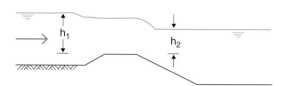

Figure 4.9. Flumes with a raised crest (Paudel, 2010).

Downstream water level above crest

distinct downstream water levels (Figure 4.9). The downstream water level is always below the crest and the downstream water level is higher than the crest; in this case h_2/h_1 should be smaller than 0.8 for modular flow.

Undershot or underflow structures

Flow under a gate, through (short) pipes or fixed plate orifices are examples of an undershot flow. In undershot flows two flow types can be distinguished (see Figure 4.10):

- free flow: the opening is relatively small ($h_1/a > 2$) and the contraction of the streamlines is significant in the vertical direction. The downstream water level (h_3) will not affect the flow and a hydraulic jump will occur downstream of the vena contracta. The discharge depends upon the gate opening, the upstream water level and the contraction;
- submerged flow: the downstream water level influences the flow and the discharge depends upon the upstream and downstream water level and the gate opening.

  The discharge through an orifice with free flow is given by Bos (1989):

$$Q = C_d C_v C_e A \sqrt{2gh} \tag{4.35}$$

where:
C_d = discharge coefficient
C_v = coefficient for velocity head
C_e = contraction coefficient, area of the vena contracta/area orifice opening
B = width of the orifice (m)
A = area of the orifice opening = $B * a$ (m^2)
h = upstream head measured from the centre of the opening to the water surface.

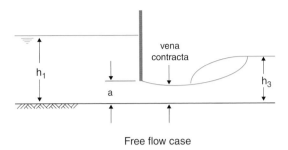

vena
contracta

h_1

a

h_3

Free flow case

h_1

a

Δh

h_2

h_3

Figure 4.10. Underflow gate
(Paudel, 2010).

Submerged flow case

The basic head-discharge equation for a submerged orifice flow is:

$$Q = C_d C_v A_{or} \sqrt{2g(h_1 - h_2)} \tag{4.36}$$

where $h_1 - h_2 =$ the head differential over the orifice. A well designed submerged orifice maintains a small approach velocity and C_v approaches unity. For a fully contracted, submerged, rectangular orifice the discharge coefficient $C_d = 0.61$ (Bos, 1989).

For a not fully contracted orifice, only a part is suppressed (bottom only or both bottom and sides) and the discharge coefficient will be:

$$C_d = 0.61(1 + 0.15r)$$

where:
$r =$ ratio of the suppressed orifice perimeter to the total orifice perimeter

If the water discharges freely through an orifice with bottom and side contraction, the flow pattern equals the flow of the free outflow underneath a vertical sluice gate and the discharge depends upon the upstream water depth and gate opening:

$$Q = C_d C_v A_{or} \sqrt{2g(y_1 - y)} = C_d C_v A_{or} \sqrt{2g(y_1 - \partial a)} \tag{4.37}$$

The variable ∂ is the contraction coefficient, the value of ∂ varies from 0.624 to 0.648. The discharge coefficient C_d is a function of h_1/a and ranges from 0.596 to 0.607 (Bos, 1989). Similarly the velocity coefficient C_v ranges from 1.0 to 1.2. Rajaratnam and Subramanya (1967) presented an alternative discharge equation:

$$Q = C_d Ba\sqrt{2gh} - C_c a \tag{4.38}$$

$$C_d = \frac{C_c}{\sqrt{1 - C_c^2 \dfrac{a^2}{h_1^2}}} \tag{4.39}$$

The contraction coefficient C_e is constant and is approximately 0.60. Estimation of C_d is accurate for higher values of h_1/a; for $h_1/a < 4$, an error of 5% might occur (Ankum, 2004).

Rajaratnam and Subramanya (1967) presented an equation for submerged flow conditions

$$Q = C_d Ba\sqrt{2g\,\Delta h} \tag{4.40}$$

where:
$\Delta h = $ head loss $(h_1 - h_3)$ (m)
$h_3 = $ downstream water depth near the gate (m)

For free flow conditions the downstream water depth should be smaller than:

$$h_3 < \frac{C_e a}{2}\left[\sqrt{1 + 8F_2^2} - 1\right] \tag{4.41}$$

where:
$C_e = $ contraction coefficient (≈ 0.60)
$F_2 = $ Froude number corresponding to the downstream water depth h_3
$h_3 = $ downstream water level above the sill (m)

For values of $a > 0.67h_1$ the discharge follows the equation for a broad-crested weir (Ranga Raju, 1981).

$$Q = 1.7BH^{3/2} \tag{4.42}$$

Pressurized flow
Culverts with a submerged (pressurized) flow can be schematized as a pipe with diameter (D) that gives the same hydraulic radius as the

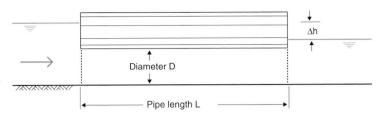

Figure 4.11. Submerged culvert or siphon (Paudel, 2010).

culvert (see Figure 4.11). The total energy loss in a pipe (culvert, siphon) is estimated as (Depeweg, 2000):

$$\Delta h = \left(f \frac{L}{D} \frac{v^2}{2g} + \Sigma \xi_1 \frac{v^2}{2g} \right) \tag{4.43}$$

The first term at the right hand side represents the friction loss and the second term the entrance, bend, exit and all the other local losses. The friction factor f depends upon the type of construction material and condition of the culvert. The value of ζ_1 can be determined from references or field measurements.

Summary
In general the depth discharge equation can be written as:

$$Q = \mu C B h \sqrt{2g \Delta h} \tag{4.44}$$

where:
  $Q =$ (discharge)
  $C =$ discharge coefficient
  $B =$ width of control section (m)
  $\Delta h =$ difference in head (m)
  $h =$ water depth (m)
  $\mu =$ coefficient for the modular limit; the coefficient for the modular limit μ is 1 for these structures with free flow conditions.

The values of the variables for different structures and flow conditions are as shown in Tables 4.8 and 4.9.
The upstream energy head (H_1) is computed from:

$$H_1 = h_1 + \frac{V_1^2}{2g} = h_1 + \frac{Q^2}{2gB^2 y_1^2} \tag{4.45}$$

For y_1 is the upstream water depth which results in:

$$y_1 = h_1 + P \tag{4.46}$$

where:
P is the crest height with reference to the upstream bed (m)

Table 4.8. Variables for different structures and flow conditions.

Structure	Flow condition	μ	h	Δh
Broad-crested	Free	1	$2/3H_1$	$H_1/3$
	Submerged	0.7	H_1	$H_1 - H_2$
Flume	Free	1	$2/3H_1$	$H_1/3$
	Submerged	0.8	H_1	$H_1 - H_2$
Short-crested	Free	1	$2/3H_1$	$H_1/3$
	Submerged	0.2	H_1	$H_1 - H_2$
Sharp-crested	Free	1	$2/3H_1$	H_1
	Submerged	1	H_1	$H_1 - H_2$
Undershot	Free	1	a	$h_1 - C_1 a$
	Submerged	1	a	$h_1 - h_2$

where:
$H_1 =$ upstream energy head above the crest (m)
$H_2 =$ downstream energy head above the crest (m)
$h_1 =$ upstream water depth (m)
$h_2 =$ downstream water depth (m)
$C_1 =$ contraction coefficient
$a =$ gate opening (m)

Table 4.9. Discharge coefficient C.

Structure	Type of structure/crest	Coefficient C	
		Minimum	Maximum
Broad-crested weir	Rounded upstream face of crest	0.930	1.02
	Sharp upstream face of crest	0.848	1.12
Flume	RBC	0.930	1.02
Short-crested weir	WES weir	0.750	
	Sharda type fall (trapezoidal crest)	0.675	
Sharp-crested weir		0.611	1.06
Undershot		0.510	0.70

Table 4.10. Variables for undershot structures.

Undershot structure	h	Δh
Free flow	a	$h_1 - C_c a$
Submerged	a	$h_1 - h_2$

where:
$h_1 =$ upstream water depth (m)
$h_2 =$ downstream water depth (m)
$C_c =$ contraction coefficient
$a =$ gate opening (m)

The same general equation is also valid for an undershot structure and the variables are given in Table 4.10; the value of *h* here is the gate opening. The discharge coefficient is included in Table 4.9.

A number of the above mentioned structures are included in the mathematical model and they are discussed in Chapter 6.

4.4.4 *General remarks*

- flow control needs extra care in networks with sediment transport. The hydraulic conditions will continuously change due to erosion or deposition in the upstream or downstream reaches. Even for the same flow and after some time, free flow conditions may switch to submerged flow conditions or vice versa. Any obstruction of the flow will change the hydraulic conditions and has a significant impact on the sediment transport capacity;
- the deposition of sediment in the upstream bed will change the bottom slope and increase both the flow velocity and the sediment transport capacity of the section. Depending upon the incoming sediment load, deposition will continue until the velocity is large enough everywhere to convey the total incoming sediment load downstream of the structure. The deposition will change the flow hydraulics and the head discharge relationship;
- a broad or a sharp-crested weir with a sill may induce sediment deposition in the upstream reach and after some time the structure may change into a sudden drop. The discharge coefficient as well as the discharge equation is not valid anymore, resulting in higher water levels, lower velocities and sediment deposition. A flume as measuring device has more advantages than a broad-crested weir in terms of sediment transport efficiency.
- if the design of the set point is low and deposition in the downstream reach occurs, then a condition may be reached, where it is not possible to maintain the flow for that set point. A proper head discharge equation requires a minimum head loss over the structure. If, due to deposition, the downstream water level rises and the available head decreases than the design discharge cannot be attained. This aspect needs special attention when designing a network with sediment transport;
- to prevent deposition upstream of a structure the water level (set point) should be more than the sum of the downstream water level and the total head loss (Δh). A set point that is too low may result in sedimentation when large discharge fluctuations occur during the irrigation season;
- to minimize sediment problems the upstream water level of a siphon should be high enough above the invert to prevent a gravity flow in the siphon;

- automatic flow control in an irrigation scheme with either upstream, downstream or volume control is based on the water level in the canal. The head discharge relationship of the canal becomes very important for a proper functioning of the automatic controls. In canals with sediment problems, the head discharge relationship will be difficult to control. The change may arise due to either deposition or erosion or to a change in the roughness during canal operation. Hence, the sediment transport aspect has to be properly analysed before deciding about the installation of any canal automation.

CHAPTER 5

Sediment Transport Concepts

5.1 INTRODUCTION

From a mathematical point of view, the interrelation between specific water flow and sediment transport conditions for a one-dimensional phenomenon without changes in the shape of the cross section can be described by the following equations (Cunge et al., 1980):

– *Continuity equation for water movement*:

$$\frac{\partial A}{\partial t} + \frac{\partial Q}{\partial x} = 0 \tag{5.1}$$

– *Dynamic equation for water movement*:

$$\frac{\partial y}{\partial x} + \frac{v^2}{C^2 R} + \frac{\partial z}{\partial x} + \frac{v}{g}\frac{\partial v}{\partial x} + \frac{1}{g}\frac{\partial v}{\partial t} = 0 \tag{5.2}$$

– *Friction factor predictor* which can be given as a function of:

$$C = f(d_{50}, v, y, S_o) \tag{5.3}$$

– *Continuity equation for sediment transport*:

$$(1-p)B\frac{\partial z}{\partial t} + \frac{\partial Q_s}{\partial x} = 0 \tag{5.4}$$

– *Sediment transport equation* which can be given as a function of:

$$Q_s = f(d_{50}, v, y, S_o) \tag{5.5}$$

where:
Q_s = sediment discharge (m³/s)
B = bottom width (m)
d_{50} = mean diameter of sediment (m)
p = porosity (dimensionless)
z = bottom level above datum (m)
v = average velocity (m/s)
y = water depth (m)
S_o = bottom slope (m/m)
R = hydraulic radius (m)

These five equations form a non-linear partial differential system, which cannot be solved analytically, but instead by a numerical method (Cunge et al., 1980). These implicit equations are not independent; they depend on each other. For instance, the water flow influences the roughness coefficient and, vice versa, the sediment transport depends heavily on the water flow.

Many mathematical models are based on the finite difference method in which the equations are replaced by a set of discrete numerical equations, which can be solved by one of the following methods:

- *Uncoupled solution*: firstly, the equations related to the water movement are separated from the total set of equations. Next, these equations are solved and the results are used to solve the sediment transport equation and the continuity equation for sediment transport. The uncoupled solution can be used for long-term simulations that experience gradual changes (Chuang et al., 1989);

- *Coupled solution*: the equations for the water movement and sediment transport are solved simultaneously, which requires general boundary conditions for the water flow and the sediment transport. In this way, numerical oscillations or instabilities are reduced. This method is recommended for short-term simulations that experience rapid changes (Chuang et al., 1989).

Although one-dimensional flow is rarely found in nature, in this book the water flow in an irrigation canal will be considered to be one-dimensional. The main assumptions for a one-dimensional flow are that the main flow direction is along the canal axis (x-direction), that the velocity is averaged over the cross section and that the water level perpendicular to the flow direction is horizontal. Other assumptions are that the effect of the boundary friction and turbulence are accounted for by the resistance laws and that the curvature of the streamlines is small and has a negligible vertical acceleration. For these assumptions, the general equations for one-dimensional flow can be described by the de Saint-Venant equations, which read as (Cunge et al., 1980):

$$\frac{\partial A}{\partial t} + \frac{\partial Q}{\partial x} = 0 \qquad\qquad \textit{Continuity equation} \qquad (5.6)$$

and

$$\frac{\partial y}{\partial x} + \frac{v^2}{C^2 R} + \frac{\partial z}{\partial x} + \frac{v}{g}\frac{\partial v}{\partial x} + \frac{1}{g}\frac{\partial v}{\partial t} = 0 \qquad \textit{Dynamic equation} \qquad (5.7)$$

The amount of water flowing into irrigation canals during the irrigation season and moreover during the lifetime of the irrigation network is not constant. Seasonal changes in crop water requirement, water supply and variation in size and type of the cropping pattern occur frequently during the life of an irrigation canal. Hence the canal must be designed and constructed with a certain degree of flexibility to deliver different amounts of water. Irrigation canals are designed for a specific design flow, but most of the time they will convey a variety of discharges and therefore it is often necessary to have a certain control to maintain the desired flow rates and required water levels.

From a computational point of view, the significance of the unsteadiness of the flow in irrigation canals must be considered from two aspects:

- Firstly, the computation of the flow at any point of the system requires a good knowledge of the response time in order to deliver the right amount of water to the right place and at the right time. The water delivery requires proper planning and operation and the flow has to be controlled. The delivery methods will result in unsteady flow conditions due to the initiation and termination of the water supply, changes in flow rate, stoppage of lateral flows, changes in gate settings, etc. Unsteady flow conditions exist most of the time and will seriously affect the water distribution. The response time of an irrigation system is a function of the distance between the origin of the disturbance and the point of interest, the celerity of the wave propagation and the operation time of the structures (see Figure 5.1). It is essential that the response time is known in view of the various changes in flow conditions (Schuurmans, 1991).

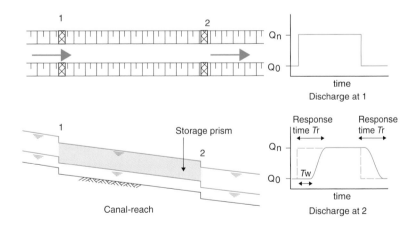

Figure 5.1. Hydro-dynamic performance of an irrigation canal with the intended and actual flow at cross section 1 and 2.

- Secondly, the flow computations for the determination of the temporal morphological changes in the canal can be based upon the assumption that the flow can be easily schematised as quasi-steady. Observations of the morphological changes in the canal bottom have shown them to be so slow that for the computation of the water movement the bottom can be considered to be fixed during a single time step (de Vries, 1965).

For the assumption as to whether the unsteady flow in the irrigation canal may be treated as a quasi-steady flow, two facts have to be considered.

- In order to avoid a very wavy water surface in the canals, which will affect the canal operation, it is recommended to maintain low Froude numbers and to limit the number to a maximum of 0.30–0.40 (Ranga Raju, 1981). For Froude numbers smaller than 0.4 the celerity of a disturbance along the water surface is not influenced by the mobility of the bed. For this condition the wave celerity is about 200 times faster than the celerity of the bed disturbances. The relative celerity of the disturbances along the water surface, being the ratio between the wave celerity and the flow velocity, is to a great extent larger than 1, while the relative celerity of the bed disturbances, being the ratio between the celerity of these disturbances and the flow velocity, is much smaller than 0.005. When the wave celerity along the water surface is much larger than the celerity of the bed disturbances it can be assumed that the disturbances of the bed will have a negligible influence on the water movement (de Vries, 1987).
- On the other hand, control structures in irrigation networks are operated in a very slow but sure way to avoid surges with steep wave fronts. This means that changes in discharge are very gradual over time and therefore the unsteady flow in an irrigation system can be approximated by a quasi-steady flow (Mahmood and Yevjevitch, 1975).

These lecture notes focus on sediment transport processes in irrigation canals and not on water delivery; therefore, the flow will be schematised as a quasi-steady flow. Hence, the terms $\delta v/\delta t$ and $\delta A/\delta t$ in the continuity and dynamic equation can be ignored. Figure 5.2 shows two hydrographs in an irrigation canal: (a) a typical one; (b) a schematised one for a quasi-steady state.

Based on these considerations the flow equations can be simplified as:
- *Continuity equation:*

$$\frac{\partial Q}{\partial x} = 0 \qquad (5.8)$$

Q is constant for steady flow
- *Dynamic equation:*

$$\frac{\mathrm{d}y}{\mathrm{d}x} = \frac{S_\mathrm{o} - S_\mathrm{f}}{1 - \mathrm{Fr}^2} \quad \text{with } \mathrm{Fr} = \frac{v}{\sqrt{gh}} \text{ for gradually varied flow} \qquad (5.9)$$

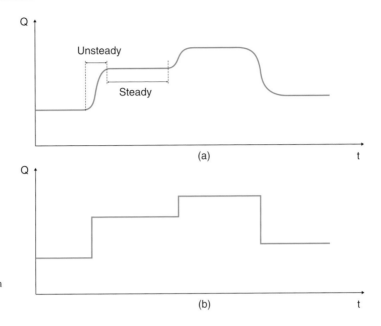

Figure 5.2. Hydrographs in an irrigation canal: (a) typical, (b) schematised.

5.2 FRICTION FACTOR PREDICTORS

5.2.1 *Introduction*

Roughness relates the mean velocity with the water depth and bottom slope and consequently influences the sediment and fluid characteristics of a canal. The correct prediction of the roughness is critical as it determines the construction costs and the type of maintenance to be expected in the future. The choice of a low roughness results in smaller cross sections and lower construction costs, but will require more maintenance to meet the design criteria. On the other hand a design roughness higher than the actual one will result in a lower water depth for the design discharge and a large number of water level regulators will be needed to raise the water level to the design level. This will increase the operational cost and change the flow pattern and the sediment transport in the canal.

Roughness of a canal can be measured under uniform flow conditions, which forms the basis for the design. Measured values are the overall roughness of the whole cross section and include the influence of various factors. Major factors that influence the roughness are (Chow, 1983):

- size and shape of the material that forms the wetted perimeter;
- vegetation on the bed and sides;
- canal surface irregularities due to localized erosion/deposition, bed forms, poor canal construction or maintenance, etc.;
- shape of the cross section and the B/y ratio;

- degree of variation in the canal alignment, the size, shape and cross section in the longitudinal direction.

The hydraulic resistance of water flowing in canals is also influenced by the development of bed forms such as ripples, mega-ripples and dunes. This resistance is measured in terms of a friction factor, for instance, the de Chézy coefficient. Other examples of friction factors are the Darcy-Weisbach and the Manning/Strickler coefficients.

The determination of the de Chézy coefficient of a movable bed is complex and requires knowledge of the implicit process of flow conditions and bed form development. The hydraulic resistance or roughness depends on the flow conditions, such as velocity, water depth and sediment transport rate, but these conditions also greatly affect the bed form development and hence the roughness. In fact, the dynamics of the bed form development and the variety of bed configurations that may simultaneously occur frustrate the development of an equation that accurately describes the roughness and the related friction factor.

The influence of the various factors will depend upon whether the canal has a rigid or erodible boundary, and carries clear water or water with sediment. Irrigation canals are manmade, so the degree of variation in canal alignment, size, shape and cross section is normally small and the effect of variation in alignment, size and shape can be ignored. For the analysis of canals carrying sediment within an erodible boundary the bed can be assumed to be free from vegetation and the effect of vegetation will be on the sides only. The irregularities in the surface may be on both the bed and sides, but the source of irregularities may be different. Irregularity due to bed forms will occur on the bed only, while irregularity due to poor construction or maintenance will be mostly on the sides. Once the canal is operated the bed will be smoothed and the only irregularities will be the bed forms. When the roughness is not uniform over the whole perimeter then the shape of the cross section and the B/y ratio will influence the overall roughness. Figure 5.3 shows the process for the determination of the effective roughness in an erodible boundary canal.

The effective roughness is not constant for different flow conditions and changes with time. Bed forms are a function of the flow and the sediment characteristics, which will change during the operation. Also the vegetation will change and there might be periodic maintenance activities during the operation. Hence a methodology should be employed to predict the roughness with time to study the performance of the design during operation of the canal system.

An important feature that determines the flow in canals is the correct estimation of the equivalent roughness. In many canals the flow very often encounters a roughness along the bottom that is different from the roughness of the sides. Some of the other characteristic conditions in irrigation canals that might influence the flow of water and sediment are

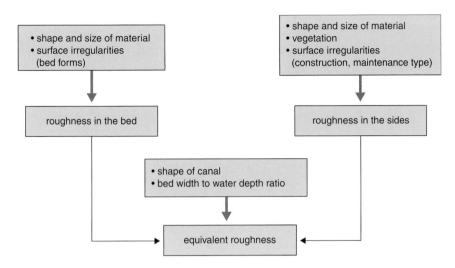

Figure 5.3. Process to derive the equivalent roughness.

the development of bed forms on the bottom, different roughness along the bottom and the sides, and vegetation on the side banks. Canals with different roughness along the perimeter need an equivalent roughness that should be based on the weighted roughness of the various components. In the past, several methods have been developed to compute the equivalent roughness from some basic assumptions concerning the flow in the canal.

5.2.2 *Bed form development*

In alluvial canals there exist two flow stages, one without any movement of bed material and the other with movement of the bed material. Canals without movement of bed material can be compared with a rigid boundary canal having an equivalent roughness height (k_s) equal to the representative bed material size (d). The resistance to flow in a movable bed consisting of sediment is mainly due to the grain roughness and form roughness. The grain roughness is generated by a skin friction force and form roughness due to forces acting on the bed forms.

Bed forms will continuously change with the flow parameters (velocity, depth) and therefore the bed roughness will also change. There are two approaches to estimate the bed roughness:

- methods based on hydraulic parameters such as mean depth, mean velocity and bed material size;
- methods based on bed form and grain-related parameters such as bed form length, height, steepness and bed-material size.

The methods based on the hydraulic parameters by Einstein and Barbarossa (1952), Engelund (1966), White et al. (1980), and Brownlie (1981b) are the most widely used ones. These methods do not explicitly account for the shape and size of bed forms in the roughness prediction. Brownlie (1981b) relates the discharge, slope and sediment characteristics with the flow depth that is used to determine the roughness of the section. White et al. (1980) gives a relation to compute the mean velocity from water depth, slope and sediment characteristics that are used to predict the friction factor. Engelund (1966) relates the mean flow velocity with the slope, water depth and a mobility parameter.

The flow in most irrigation canals is sub-critical, meaning that the Froude number is smaller than 1. More especially the flow in irrigation canals is found in the lower flow region with Froude numbers smaller than 0.7 and even smaller than 0.4 (Ranga Raju, 1981). Bed features for these low flow regimes can be described as a flat bed, as ripples or as dunes. The latter two forms are characterized by rough triangles in the longitudinal profile with a gentle sloping upstream face and a more inclined downstream face (see Figure 5.4). No sharp distinction exists between ripples and dunes, but they show some subtle differences. Engelund (1966) described these bed forms as:

- *flat bed*: a flat bed is a surface without any bed form. There is no motion of the sediment on flat beds;
- *ripples*: ripples are a bed form to be found in canals with bed material smaller than 0.6 mm; the wave length is shorter than 30 cm and the wave height is a few centimetres. When the velocity is slightly greater than the threshold value several ripples will be formed in the bed. The ripple geometry is almost independent of the flow conditions;
- *dunes*: dunes are a bed form that occurs in flows with a larger velocity. Dunes develop for all sizes of bed sediment and their length and height are greater than those of ripples. Moreover, the sediment transport in canals with dunes is larger than in canals with ripples. The geometry of the dunes strongly depends on the flow depth.

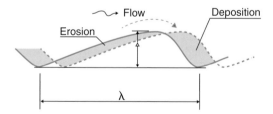

Figure 5.4. Schematic representation of bed forms for the low flow regime.

Several authors have analysed the bed forms as observed under flume and field conditions and tried to explain the type of bed forms for certain flow conditions. They presented graphical solutions for the prediction of bed forms by using dimensional and non-dimensional plots. Some of these authors are Liu (1957), Simons and Richardsons (1966), Bogardi (1974) and van Rijn (1993). Each theory is based on a particular classification parameter as presented in Table 5.1.

Table 5.1. Classification parameter used in bed form theories.

Author	Classification parameter
Liu (1957)	u_*/w_s and $u_* d_{50}/v$
Simons & Richardson (1966)	$\tau^* v$ and d_{50}
Bogardi (1974)	$g d_{50}/u_*^2$
van Rijn (1993)	T and D_*

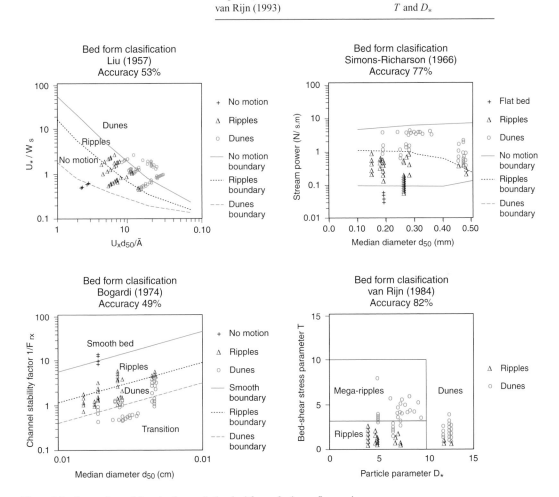

Figure 5.5. Comparison of theories for predicting bed forms for lower flow regimes.

In order to find an appropriate theory to describe and predict the type of bed forms in irrigation canals the theories developed by Liu (1957), Simons and Richardsons (1966), Bogardi (1974) and van Rijn (1984c) have been compared on the basis of field and laboratory data (Figure 5.5). These theories explain the bed forms for one-directional flows and for homogeneous conditions, both in time and space (Jansen, 1994). Brownlie

(1981a) compiled an extensive set of data on sediment transport and bed forms from the laboratory and field data available at that time. For the development of the computer program SETRIC (see Chapter 6), a selection of data was made from the compilation that was published in 1981 and this data was used to compare the various theories on bed forms.

The selection of the data was based on the flow conditions and the sediment characteristics that are normally encountered in irrigation canals (Méndez, 1995 and Paudel, 2010). The selection criteria included:
- the compiled data had to comprise all the quantities necessary to compute the four classification parameters as mentioned in Table 5.1;
- the sediment size d_{50} had to be smaller than 0.5 mm;
- the Froude number had to be smaller than 0.5;
- the shear stress on the bottom had to be smaller than 5 N/m^2;
- the B/h ratio had to be larger than 10 in order to minimize the influence of the sides;
- the compiled data had to include a detailed description of the bed forms.

The theories were compared on a relative basis, which means that the number of relatively accurately predicted (*well-predicted*) bed forms according to each theory was related to the total number of observed bed forms. The predictability of each of the four theories was measured in terms of accuracy (number of *well-predicted* values), which is represented by:

$$\frac{\text{measured value}}{f} \leq \text{predicted value} \leq \text{measured value} * f \tag{5.10}$$

$$\text{accuracy (\%)} = \frac{\text{number of well-predicted values}}{\text{total values}} * 100 \tag{5.11}$$

The accuracy of each method to predict the bed forms has been used to draw some conclusions:
- the theories presented by van Rijn (1984c) and Simons and Richardson (1966) appear to be the best theories to predict the bed form in irrigation canals. The accuracy shows that approximately 82% and 77% respectively of the observed bed forms are predicted relatively accurately (*well predicted*) by the two theories;
- all the bed forms described for the lower flow regime (ripples, mega ripples and dunes) can be expected in irrigation canals.

According to van Rijn (1993), bed forms can be classified by the following parameters:
– particle parameter D_*, which reflects the influence of density and diameter of the particle, gravity and viscosity:

$$D_* = \left[\frac{(s-1)g}{\nu^2} \right]^{1/3} d_{50} \tag{5.12}$$

Characteristic values for the particle parameter D_* in irrigation canals are in the range of 1.5 to 7.3 (Table 5.2).

Table 5.2. D_* parameter for sediment sizes encountered in irrigation canals.

d_{50} (mm)	0.05	0.1	0.15	0.20	0.25	0.30	0.35	0.40	0.45	0.50
D_*	1.2	2.5	3.7	5.0	6.2	7.5	8.7	10.0	11.2	12.5

– Excess bed shear stress parameter T which is defined by:

$$T = \frac{\tau' - \tau_{cr}}{\tau_{cr}} \tag{5.13}$$

with

$$\tau' = \rho g \left(\frac{v}{C'}\right)^2 \quad \text{and} \quad C' = 18 \log\left(\frac{12h}{4.5 d_{90}}\right) \tag{5.14}$$

The ranges described by van Rijn (1993) are shown in Table 5.3.

Table 5.3. Classification of bed forms according to van Rijn (1984c).

Transport regime		Particle parameter	
		$1 \leq D_* \leq 10$	$D_* > 10$
Lower	$0 \leq T \leq 3$	Mini ripples	Dunes
	$3 \leq T \leq 10$	Mega ripples	Dunes
	$10 \leq T \leq 15$	Dunes	Dunes

Figure 5.6 shows the ranges established by van Rijn (1984c) to distinguish ripples, mega ripples and dunes. The curves of the maximum values of the classification parameter T for small irrigation canals (hydraulic radius $R = 0.5$ m) are given for Froude numbers of 0.15, 0.25, 0.35 and 0.45. For larger irrigation canals ($R \approx 4$ m), the curves are for Froude numbers of 0.08, 0.12 and 0.16. Higher Froude numbers in large canals show that the actual shear stress on the bottom is larger than the approximate 4–5 N/m^2 allowed.

5.2.3 *Effects of grains and bed forms on the roughness of the bed*

Methods based on bed form parameters separate the total roughness into that due to grains and bed forms. The methods require the shape and size of bed forms to determine explicitly the equivalent roughness height. Méndez (1998) tested the accuracy of the friction factor predictors with selected data and concludes that the method by van Rijn (1984b) based on bed form parameters gives the best results.

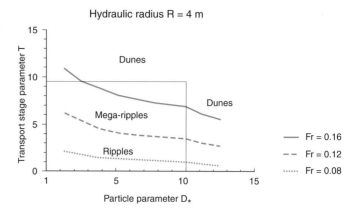

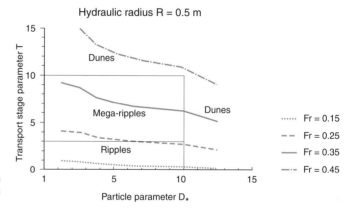

Figure 5.6. Classification of bed forms according to van Rijn (1984c) and expected bed forms in irrigation canals.

The total effective shear stress (τ) can be separated into two parts, namely that due to skin resistance τ' and form resistance τ'':

$$\tau = \tau' + \tau'' \tag{5.15}$$

Similar to the separation of shear stress in grain and form resistance, the Nikuradse's equivalent sand roughness height k_{se} can be separated as (van Rijn, 1982):

$$k_{se} = k_s' + k_s'' \tag{5.16}$$

where
k_s' = roughness due to grain (m)
k_s'' = roughness due to bed form development (m)

This equation gives the total equivalent roughness due to the grain size of the bed material and the bed form created due to movement of the particles. There are two possible stages for which the equivalent roughness needs to be investigated; one is without any movement of the particles (plane bed) and the other with movement of the particles (bed forms).

Van Rijn (1982) proposes the roughness due to the grains as: $k'_s = 3d_{90}$ and assuming a regular sediment size distribution $k'_s = 4.5d_{50}$ for $d_{90} = 1.5d_{50}$.

When the velocity increases the bed material starts moving and the bed feature changes. According to van Rijn (1982) the term k''_s is related to the bed form height, bed form steepness and bed form shape. The following relationship is given for k''_s:

$$k''_s = f(\Delta, \Delta/\lambda, \gamma) \tag{5.17}$$

for ripples:

$$k''_s = 20\gamma_r \Delta_r \left(\frac{\Delta_r}{\lambda_r}\right) \tag{5.18}$$

for dunes:

$$k''_s = 1.1\gamma_d \Delta_d(1 - e^{-25\Delta_d/\lambda_d}) \tag{5.19}$$

where:
Δ_r = ripple height = 50 to 200 * d_{50} (m)
γ_r = ripple presence ($\gamma_r = 1$ for ripples only)
λ_r = ripple length = 500 to $1000d_{50}$ (m)
γ_d = form factor = 0.7 for field conditions and 1.0 for laboratory
   conditions
Δ_d = dune height (m)
λ_d = dune length (m) = 7.3 * h

5.2.4 Roughness of the side slopes

The roughness height due to the surface material is related to the median particle diameter (d_{50}) as $k'_s = 4.5d_{50}$. This value is recommended for a rigid boundary, but for a movable bed of loose material the roughness will be higher, because a perfectly plane side slope will not exist in natural conditions and small irregularities will always be present. Van Rijn (1993) suggests a minimum value of $k'_s = 0.01$ m. Moreover the Manning's roughness coefficient for an ideally finished earthen canal is 0.016 to 0.020 (Chow, 1983). This coefficient (n) is an average value that is assumed to be uniform over the whole cross section. The value of n can be related to de Chézy's roughness coefficient for the side part only by:

$$C_s = \frac{R_s^{1/6}}{n} \tag{5.20}$$

where:
C_s = Chézy's roughness coefficient on the side slope only ($m^{1/2}$/s)
R_s = hydraulic mean radius of the sides (m)

Next the de Chézy's roughness coefficient can be related to the roughness height of the sides as:

$$C_s = 18 \log\left(\frac{12R_s}{k_{ss}}\right) \tag{5.21}$$

Roughness caused by surface irregularity may be due to the type of material, construction methods and workmanship, ageing of the side slope, rain cuts, slides of the banks, etc. Chow (1983) gives a classification of the surface irregularities and correction factors for Manning's coefficient for four categories:

- *Ideal*: refers to the best attainable surface for the construction material. Newly constructed or well-maintained canals with perfect workmanship belong to the ideal type of canal. No correction is needed for this type of surface. For earthen canals the value of Manning's n is 0.018;
- *Good*: refers to newly constructed or weathered but well maintained canals with good to moderate finishing. A value of 0.005 is added to the value of Manning: $n = 0.023$;
- *Fair*: the surface of the canals that are moderately to poorly excavated and also includes the canals that have been excavated by machines and have eroded side slopes. A value of 0.01 is added ($n = 0.028$);
- *Poor*: badly eroded or sloughed side slopes, large rain cuts and excavations not in the proper shape. A value of 0.02 is added ($n = 0.038$).

Vegetation reduces the effective flow area and increases the roughness (see Table 5.4). Vegetation growth is more pronounced in clear water; however, the nutrients in water with sediment may help the growth of weeds. The degree of obstruction by vegetation is highly variable and depends upon the type, height, density and flexibility of the vegetation, submerged or un-submerged conditions, water level, and flow velocity (Paudel, 2010). Kouwen (1988) gives an empirical relation to calculate the equivalent roughness height for the given vegetation:

$$k_s = 0.14 h_g \left[\frac{(mei/\tau)^{0.25}}{h_g}\right]^{1.59} \tag{5.22}$$

Table 5.4. Weed factor for different types of vegetation at full growth (derived from Chow's (1983) suggested n for vegetation).

Category	Description	Weed factor (Chow)
Low	– dense growth of flexible turf grass ($h/h_g = 2$–3) – supple seedling tree switches ($h/h_g = 3$–4)	1.25–1.5
Medium	– turf grasses ($h/h_g = 1$–2) – stemmy grasses, weeds or tree seedlings ($h/h_g = 2$–3) – brushy growth, moderately dense	1.5–2.5
High	– turf grasses ($h/h_g = 1$) – willow or cottonwood trees 8–10 years old – bushy willows	2.5–3.5
Very high	– turf grass ($h/h_g = 0.5$)	3.5–6.0

h = water depth and h_g = height of vegetation.

mei is a parameter that is a function of stem density, elasticity modulus and second moment of area and is:

for green grass: $mei = 319h_g^{3.3}$

for dead grass: $mei = 25.4h_g^{2.26}$

where:

h_g = height of the vegetation (m)

k_s = equivalent roughness height (m)

τ = local shear stress (N/m^2)

Another simple way to incorporate the effect of weed is given by Chow (1983), who has suggested a classification of vegetation and a correction to be added to the Manning's n.

Concerning the weed effect it can be mentioned that the analysis of the sediment transport process assumes that the canal bed is free of weeds; however, the side slopes may have weed effects. In the case that weed grows on the side slope then the roughness due to surface irregularities is neglected. It is assumed that the vegetation fully covers the surface irregularities. The equivalent roughness of a canal with bed forms on the bottom and vegetation on the side slopes depends on the degree of obstruction by the weed growth. Querner (1993) presents typical variation of the relative obstruction and elative roughness coefficient during the weed growth period (see Figures 5.7 and 5.8). This variation gives only tendencies for the changes in relative obstruction and roughness.

To obtain the actual roughness the friction factor has to be increased by a weed factor F_w. The weed factor depends on local flow conditions, vegetation characteristics (growing period, type etc.). Méndez (1998) mentions values of F_w between 0.1 (a very densely overgrown canal) and 1 (an ideally cleaned canal). Table 5.5 shows some values of the weed factor F_w, which is based on the data from Nitschke (1983). These values are only applicable for the specific flow conditions and vegetation described in that research.

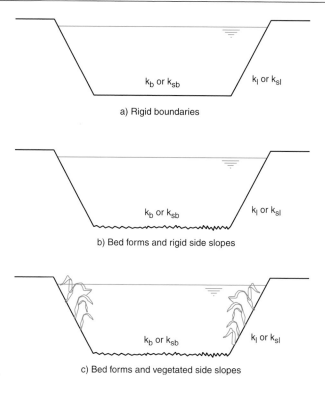

Figure 5.7. Types of equivalent roughness in trapezoidal canals.

a) Rigid boundaries

b) Bed forms and rigid side slopes

c) Bed forms and vegetated side slopes

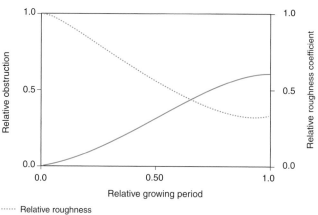

Figure 5.8. Variation in the relative obstruction and relative roughness during the weed growth period (Querner, 1993).

······ Relative roughness
── Relative obstruction

Depending upon the type of weed and relative height of weed compared to the water depth the maximum weed factor is determined. It is assumed that the influence on the roughness will increase linearly from the start to the full growth. Canal maintenance during the irrigation season

Table 5.5. Weed factor for different obstruction degrees (Nitschke, 1983).

Obstruction degree	Weed factor F_w
<5%	1.0
5–10%	0.9
10–15%	0.8
15–25%	0.7
25–35%	0.5
35–50%	0.4
50–75%	0.2
>75%	0.1

may reduce the full effect of the vegetation. The duration and frequency of maintenance is necessary to determine the weed factor.

The weed effect is time dependent and can be predicted to some degree, since the type and other properties of vegetation can be easily observed. The variation of the weed factor during a certain period will depend on the type of maintenance (see Table 5.6 and Figure 5.9). Three scenarios with different maintenance policies will be considered:
– *no maintenance*: no weed clearing
– *well maintained*: weed clearing when the degree of obstruction is 25% (approx. 2 months)
– *ideally maintained*: continuous weed clearing.

Table 5.6. Weed factors according to the type of maintenance.

Type of maintenance	Weed factor	
	Maximum	Minimum
No maintenance	1.0	0.1
Well maintained	1.0	0.7
Ideally maintained	1.0	1.0

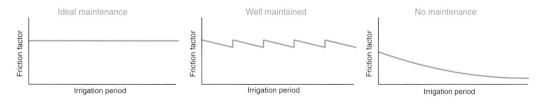

Figure 5.9. Three possible maintenance scenarios.

The weed factor is defined as:

$$\text{weed factor} = \frac{\text{roughness including weed effect}}{\text{roughness without weed effect}} \tag{5.23}$$

The actual roughness can be calculated by using the weed factor: Actual roughness $= F_w *$ Initial roughness.

Next the equivalent roughness for a cross section with obstruction can be estimated by computing the initial roughness of the whole cross section, depending on whether there is either a single roughness or an equivalent roughness or based on the type of roughness of the bottom and sides. Next the actual equivalent roughness is reduced by the weed factor F_w.

5.2.5 *Equivalent roughness for non-wide irrigation canals*

The most common cross section of an irrigation canal has a trapezoidal shape with a relatively small bottom width-water depth ratio. In this type of cross section, the velocity distribution is heavily affected by the varying water depth perpendicular to the sidewall and by the boundary condition imposed on the velocity by the sidewall. An important interaction and transfer of momentum between the side parts and the central part of the canal will take place. The existing methods to estimate the composite roughness were developed for rivers in which the cross section is divided into a main canal and two flood plains. The assumptions made are not valid for non-wide, trapezoidal or rectangular canals. The main shortcomings are firstly, the lack of attention to the effect of the varying water depth on the friction and secondly, the hypothesis that the mean velocity and/or the hydraulic radius is the same in all the subsections.

Normally, the flow in irrigation canals is turbulent and it can be assumed that the lateral velocity distribution in a trapezoidal canal is more governed by the varying water depth above the sides than by the boundary condition imposed by the wall. Based on this hypothesis, the composite roughness can be derived by assuming that the cross section consists of an infinite number of stream tubes (slices). The water depth and the local friction of each stream tube govern the resistance in the stream tube. To evaluate the resistance it is assumed that there is no transfer of momentum between the stream tubes and that the local resistance can be expressed by the de Chézy coefficient (see Figure 5.10).

In irrigation canals the roughness of the bottom and the sides are often different. Several cases of composite roughness in an irrigation canal can be distinguished:

- *Rigid boundaries*: When the boundary is rigid and unchanging then the resistance only depends on the skin roughness. The composite roughness depends on the material of the walls and bottom. Roughness

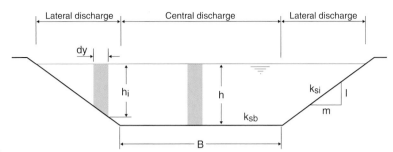

Figure 5.10. Composite roughness in a trapezoidal canal.

coefficients for the walls and bottom follow from recommended values for the type of material on the bottom and walls, respectively. Chow (1983) gives an extensive list of roughness coefficients.

- *Movable bed*: When the canal bed is movable then the roughness may change through the development of bed forms on the bottom, but not on the sides. The sediment particles on the sides are subject to the fluid force and to the gravity force. Therefore, the critical shear stress for the initiation of motion of these particles is much smaller due to the component of the gravity force along the slope. For slopes that are steeper than the angle of repose the critical shear stress for the initiation of motion is reduced to zero (Ikeda, 1982b). Recommended values of *m* for unlined irrigation canals are between 1 and 3. For most cases the slope angle is between 18.5° and 45° and exceeds the natural angle of repose of wet sand, which is between 15° and 25° (Kinori, 1970). Due to this fact, it is expected that sediment particles will be deposited on the bottom and not on the slopes.

The two most common cases for composite roughness of canals with a movable bed are:

- *Bed forms on the bottom and flat sidewalls*: the roughness of the bottom follows from the bed form characteristics and the roughness of the sidewall follows from the type of material.
- *Bed forms on the bottom and vegetation on the sidewalls*: vegetation on the sidewalls is a type of roughness (Chow, 1983), which is often given by a single value drawn from field measurements. However, the roughness of canals with vegetation is more complex and is a function of many variables related to the flow conditions and vegetation characteristics, which cannot be expressed by a single value. The flow resistance for a canal with vegetation is difficult to determine and requires ample research before the phenomena involved are completely understood (Kouwen, 1969 and 1992). So far, it is almost impossible to make a proper estimate of the resistance based on only analytical or theoretical considerations only (Querner, 1993).

Irrigation canals are normally not very wide (B/y ratio < 8), hence the influence of the roughness of the side slope will be significant in the overall roughness value. Different methods are available to determine the equivalent (or composite) roughness (k_{se}) of a section. Yen (2002) discussed in detail the different aspects of computing equivalent roughness of an open canal.

Method 1. Vanoni (1975), Chow (1983), and Raudkivi (1990) have stated that the equivalent roughness of a cross section can be found by considering the total cross section as an area composed by a number of subsections. The mean velocity and energy gradient of each subsection is the same as the velocity and energy gradient of the entire cross section. The equivalent roughness for the whole cross section is then determined by:

$$A = \sum_{i=1}^{N} A_i \tag{5.24}$$

$$\left(\frac{vP^{2/3} n_e}{S^{1/2}} \right)^{3/2} = \sum_{i=1}^{N} \left(\frac{vP_i^{2/3} n_i}{S^{1/2}} \right)^{3/2} \tag{5.25}$$

This was proposed independently by Horton and by Einstein (Chow, 1983).

$$n_e = \left(\sum_{i=1}^{N} \frac{P_i n_i^{3/2}}{P} \right)^{2/3}$$

Method 2. This method was proposed by Pavlovskiĭ, by Muhlhofer, and by Einstein and Banks (Chow, 1983). Chow (1983), Krishnamurthy and Christensen (1972), and Motayed and Krishnamurthy (1980) assumed that the total hydraulic resistance in a cross section is equal to the summation of the flow-resisting forces in each subsection and that the hydraulic radius of each subsection is equal to the radius of the whole cross section. The total force resisting the flow is equal to the sum of the forces resisting the flow in each subsection.

$$\tau P = \sum_{i=1}^{N} \tau_i P_i \quad \text{with } \tau = \rho g R S \quad \text{and} \quad S = \frac{v^2 n^2}{R^{4/3}} \tag{5.26}$$

$$n_e = \left(\sum_{i=1}^{N} \frac{P_i n_i^2}{P} \right)^{1/2}$$

Method 3. Ida (1960) derived a relation by equating the discharge through all the subsections with the whole section. Asano et al. (1985) modified the method by replacing the hydraulic radius R by an equivalent hydraulic radius R_e, which is given by:

$$R_e = \left(\sum_{i=1}^{N} \frac{P_i R_i^{5/3}}{P} \right)^{3/5} \tag{5.27}$$

$$n_e = \frac{\sum\limits_{i=1}^{N} P_i R_i^{5/3}}{\sum\limits_{i=1}^{N} \dfrac{P_i R_i^{5/3}}{n_i}} \tag{5.28}$$

Method 4. Krishnamurthy and Christensen (1972) proposed that the summation of the discharges in subsections with roughness coefficient $k_{s,i}$ (i subscript for the subsection) is equal to the summation of the discharges of all the subsections with an equivalent roughness k_{se}. The flow in each section is assumed to be turbulent and the velocity distribution is described by the logarithmic law. The hydraulic roughness is related to Manning's roughness coefficient by (Henderson, 1966):

$$n = 0.034 k_s^{1/6} \tag{5.29}$$

The equivalent roughness value is given by:

$$\ln n_e = \frac{\sum\limits_{i=1}^{N} P_i R_i^{3/2} \ln n_i}{\sum\limits_{i=1}^{N} P_i R_i^{3/2}} \tag{5.30}$$

where:
$n_e =$ equivalent Manning's roughness coefficient for the whole cross section;
$n_i =$ Manning's roughness coefficient in subsection i;
$u_{*i} =$ shear velocity in subsection i;
$v_i =$ mean velocity in subsection i;
$A =$ area of the entire cross section;
$A_i =$ area of subsection i;
$S =$ energy gradient;
$q_i =$ discharge in subsection i;
$\tau_e =$ shear stress for the whole cross section;
$\tau_i =$ shear stress in subsection i;
$P =$ wetted perimeter of the whole cross section;
$P_i =$ wetted perimeter of the subsection i;
$R =$ hydraulic radius for the whole cross section;

R_i = hydraulic radius for the subsection i;
R_e = equivalent hydraulic radius;
h_i = water depth of subsection i;
k_{si} = hydraulic roughness in each subsection i;
k_{se} = equivalent or equivalent hydraulic roughness;
dy = width of subsection i;
N = total number of subsections.

Method 5. Méndez (1998) proposed to divide the discharge into discharges through the central part and side slope parts and the total discharge is the sum of subsection discharges. $Q = Q_{cen} + Q_{lat}$ (see Figure 5.11). Neglecting the effect of the momentum transfer, the discharge through a stream column of width dy and water depth h_i, with a local roughness height k_{si} can be calculated as:

$$Q_i = C_i h_i \, dy \sqrt{h_i S_f} \tag{5.31}$$

where

$$C_i = 18 \log \frac{12 h_i}{k_{si}} \tag{5.32}$$

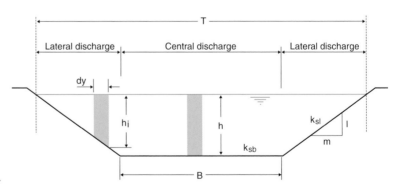

Figure 5.11.   Schematization of flow through a trapezoidal canal.

The discharge through each of the subsections is the summation of the flow in each stream tube and can be written as:

$$Q = Q_{lat} + Q_{cen} = 2 \int_0^{mh} C_i h_i \sqrt{h_i S_f} \, dy + \int_0^B C_i h \sqrt{h S_f} \, dy \tag{5.33}$$

Solving the equation gives:

$$Q = \frac{18 S_f^{1/2} h^{3/2}}{2.3} \left[\frac{4}{5} mh \left(\ln \frac{12h}{k_{sl}} - \frac{2}{5} \right) + B \ln \frac{12h}{k_{sb}} \right]$$

If the roughness height in the bed (k_{sb}) and sides (k_{sl}) is the same and is equal to k_{se} then the equation can be written as:

$$Q = \frac{18 S_f^{1/2} h^{3/2}}{2.3} \left[\frac{4}{5} mh \left(\ln \frac{12h}{k_{se}} - \frac{2}{5} \right) + B \ln \frac{12h}{k_{se}} \right] \qquad (5.34)$$

Comparing the equations gives:

$$\ln k_{se} = \frac{0.8 \ln k_{sl} + (B/h) \ln k_{sb}}{0.8m + (B/h)} \qquad (5.35)$$

$$C_e = 18 \log \left(\frac{12R}{k_{se}} \right) \qquad (5.36)$$

Since the lateral transfer of momentum and its effect on the velocity distribution across the canal is not included, the equation will over predict the flow rate. If this equation is to be used for discharge calculation then a correction factor for the velocity distribution has to be applied. The modified effective Chézy's roughness coefficient is then $C_é = f_e C_e$, with f_e being the correction factor for the effective Chézy coefficient, which is a function of the B/y-ratio, side slope m and roughness on the bed and side slopes (Méndez, 1998).

Method 6. Instead of applying a correction on the basis of the velocity distribution the equivalent roughness is computed using the previous equation only. Comparing the equations gives:

$$\ln k_{se} = \frac{0.8 \ln k_{sl} + (B/h) \ln k_{sb}}{0.8m + (B/h)} \qquad (5.37)$$

$$C_e = 18 \log \left(\frac{12R}{k_{se}} \right) \qquad (5.38)$$

where:
  $m =$ side slope;
  $h =$ water depth in m;
  $B =$ bottom width in m;
  $k_{se} =$ hydraulic roughness in each stream tube i;
  $k_{sl} =$ hydraulic roughness along the sides;
  $k_{sb} =$ hydraulic roughness along the bottom.

5.2.6 Comparison of the equivalent roughness predictors in trapezoidal canals

All the 6 methods for the prediction of the roughness were compared using the Krüger (1988) data set. Considering the general conditions of irrigation canals, the following criteria have been used for selecting the data:
– trapezoidal cross section;
– Froude number less than 0.5;
– *B/y* ratio less than 8;
– the ratio of equivalent roughness height on the bed and the sidewall (k_{sb}/k_{se}) in the range of 50 to 0.02.

A total of 19 records were selected from the compilation of Krüger. Table 5.7 shows the summary of the selected data.

Table 5.7. Characteristics of selected data.

Test	*B/y* ratio	Side slope	k_{sl} (mm)	k_{sb} (mm)	k_{sb}/k_{sl}	No. of records
1	3.0–5.8	1	0.054	1.047	19.800	3
2	1.8–5.7	1	8.400	1.047	0.125	8
3	3.9–7.9	2	0.054	1.047	19.800	8

The calculation process used in the comparison of the selected data comprises the following steps:
– since it is not possible to measure Manning's roughness coefficient separately for the bed and side slopes, the representative size of the particles (d_{50} or d_{90}) was used to determine the equivalent roughness height (k_s). Various authors have given different estimation methods. Henderson (1966) suggests 2~3 * d_{50}, van Rijn (1982) suggests 1~3 * d_{90}, while Krüger (1988) has suggested using d_{90} for a plane bed. For the calculations $k_s = d_{90}$ has been used;
– the local Manning's roughness coefficients (n_b, n_i) is determined from (Henderson, 1966):
$$n = 0.031(3.28k_s)^{1/6}$$
the flow can be divided into hydraulically smooth, transition or rough depending on the following conditions (van Rijn, 1993):
smooth flow:

$$\frac{u_* k_s}{\nu} \leq 5 \qquad\qquad (5.39)$$

transition flow:

$$5 < \frac{u_* k_s}{\nu} < 70 \qquad\qquad (5.40)$$

rough flow:

$$\frac{u_* k_s}{v} \geq 70 \tag{5.41}$$

- according to the mentioned criteria, the flow for the selected data set is defined as a transition flow; hence the velocity distribution will be affected by viscosity as well as by the bed and side roughness. The de Chézy's roughness coefficient for a transition flow is given by:

$$C = 18 \log\left(\frac{12R}{k_s + 3.3v/u_*}\right) \tag{5.42}$$

- next the average value of the Manning's roughness coefficient is computed from the k_s value from the de Chézy equation and is compared with the derived n using different methods.

The comparison between the 6 methods is made on the basis of the number of well-predicted values within an error band. If K is the error factor then the error band is the range between the measured value/K and measured value $* K$. Once a predicted value lies within the above given band then it will be seen as a well-predicted value. The accuracy of a method is given by:

$$\text{accuracy} \ (\%) = \frac{\text{number of well predicted values}}{\text{number of total values}} * 100\% \tag{5.43}$$

The accuracy of the predictability of the different methods is given in Figure 5.12 for different error factors. The result shows that the prediction method 6 gives the best results for the equivalent roughness in a trapezoidal canal section. Hence, method 6 will be used for the evaluation of the overall roughness of a canal section.

The comparison of the methods to predict the equivalent roughness in a trapezoidal cross section with data from Krüger results in the following conclusions:

- comparison of the overall performance of the methods with the experimental data set shows that the methods can be ranked in descending order, namely 6, 5, 2, 1, 4 and 3;
- methods 1, 2, 3 and 4 were developed for river conditions, in which the channel consists of a main canal and two parallel flood plains. The water depths in both main canal and flood plains do not vary in a lateral direction;
- method 6 appears to give better results for error factors smaller than 1.2. This method predicts the measured data with an accuracy higher than 90% for an error factor of 1.15. The minimum standard error of the predicted values was also observed for method 6;
- methods 1 and 2 behave similarly. The assumptions for both methods give similar results; they weigh the side parts in the same way as the

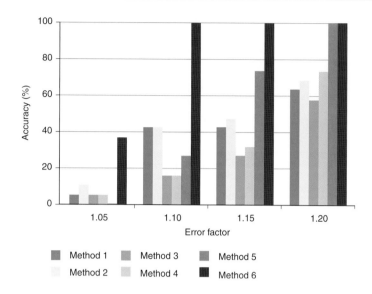

Figure 5.12. Comparison of well-predicted values for different values of error factor using Krüger's data.

central part without considering the differences in velocity or water depth;

- methods 3 and 4 are fairly similar; they differ in the description of the mean velocity in the sub-sections.

5.2.7 Prediction of composite roughness in a rectangular canal

For canals with a rectangular cross section, the existing methods to find the equivalent roughness cannot be directly used. Rectangular cross sections do not have a clearly defined area that can be associated with each type of roughness along the wetted perimeter. Therefore, the estimate of the equivalent roughness follows the same principles as used for the sidewall correction method, which is a calculation procedure initially proposed by Einstein (1942) to determine the shear stress at the bottom as well as the shear velocity, friction factor, etc. The method does not include any correction for the effect of the sidewalls on the velocity distribution or sediment transport (Ranga Rayu et al., 1977).

Méndez described a method for a non-wide rectangular canal with bottom width B and water depth h (Figure 5.13). The rectangular canal is replaced by another canal, namely a *wide canal* with a bottom width B_* ($B_* = B + 2h$) and a water depth R ($R = A/P$). The area $A = R*(B + 2h)$ and the total discharge of the *wide,* rectangular canal is expressed by:

$$Q = vA = 2C_L h R\sqrt{RS_f} + C_b BR\sqrt{RS_f} \qquad (5.44)$$

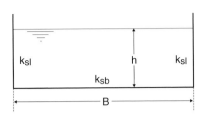

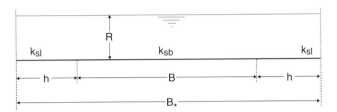

Figure 5.13. Non-wide
rectangular canal schematised
as a wide canal.

Replacing the de Chézy coefficients in the equation (5.44) by a function
of the surface roughness and water depth gives:

$$Q = 2\left(18\log\frac{12R}{k_{sl}}\right)hR\sqrt{RS_f} + \left(18\log\frac{12R}{k_{sb}}\right)BR\sqrt{RS_f} \qquad (5.45)$$

Expressing the discharge in terms of the equivalent surface roughness
gives:

$$Q = \left(18\log\frac{12R}{k_{se}}\right)(B+2h)R\sqrt{RS_f} \qquad (5.46)$$

Equating both equations and rearranging the terms gives:

$$\log k_{se} = \frac{2\log k_{sl} + (B/h)\log k_{sb}}{2 + (B/h)} \qquad (5.47)$$

Expressed in terms of the Manning's coefficient results in:

$$n_e = \frac{2 + \dfrac{B}{h}}{\dfrac{2}{n_l} + \dfrac{B}{h}\dfrac{1}{n_b}} \qquad (5.48)$$

Méndez (1998) has tested this method for rectangular canals with a
selected set of laboratory data from Krüger (1988). The selection criteria for the data were similar to those used for the trapezoidal canals
and the comparison of the prediction method for the equivalent roughness

in rectangular canals with the selected data is similar to the one given earlier.

5.2.8 *Effect of bed forms on the flow resistance*

The hydraulic resistance to flowing water in open channels is affected by the development of bed forms such as ripples, mega ripples and dunes. The hydraulic resistance due to the bed roughness is expressed by a friction factor. Most common friction factors are:
 − the Darcy-Weisbach friction factor f:

$$f = \frac{8gRS_f}{v^2} \text{ from } f = \frac{8g}{C^2} \tag{5.49}$$

 − the de Chézy coefficient:

$$C = \frac{v}{\sqrt{S_f R}} \tag{5.50}$$

 − the Manning (n) or Strickler (k) roughness coefficient:

$$n = \frac{R^{2/3} S_f^{1/2}}{v} \quad \text{for Manning} \tag{5.51}$$

$$k = \frac{v}{R^{2/3} S_f^{1/2}} \quad \text{for Strickler with } n = 1/k \tag{5.52}$$

For the same conditions the Manning coefficient can be related to the de Chézy coefficient by:

$$C = \frac{R^{1/6}}{n} \tag{5.53}$$

In these lecture notes, mainly the de Chézy coefficient will be used to describe the friction in irrigation canals. The use of the Darcy-Weisbach and Manning/Strickler roughness coefficients follows from the equations described above.

Not only the wall (grain) roughness, but also the bed forms, are elements that resist the flow and the total resistance in channels with a movable bed consists of two components:
 • the surface or skin resistance due to the grain roughness; the resistance has a grain-related shear stress τ':
 • the form resistance due to hydrodynamic forces acting on the bed forms has a form-related shear stress τ''.

Figure 5.14 presents the total shear stress due to skin and form resistance for different velocities. For low velocities, the bed shear stress is smaller than the threshold value and no motion of particles occurs: the bed remains flat and the total bed shear stress is represented by the grain related shear stress τ'. For increasing velocity, especially beyond the threshold, transport of sediment will start, the bed becomes unstable and some bed

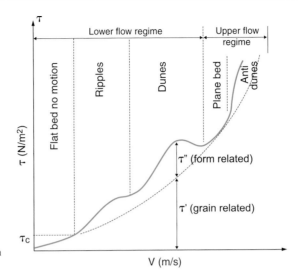

Figure 5.14. Total shear stress due to skin resistance and bed forms as a function of the mean velocity (Jansen, 1994).

configuration takes place. For velocities that are slightly larger then the threshold value small disturbances in the form of ripples will develop. Higher velocities will produce dunes and at this stage, the total bed shear stress consists of the two above-mentioned components, namely the skin and the form resistance. The total shear stress is:

$$\tau = \tau' + \tau'' \tag{5.54}$$

The definition of the Darcy-Weisbach friction factor gives:

$$\tau = \rho g R S \tag{5.55}$$

$$\tau = \frac{f \rho v^2}{8} \tag{5.56}$$

$$f = \frac{8 g R S}{v^2} \tag{5.57}$$

In the same way, the total friction factor f can be considered to have two components:

$$f = f' + f'' \tag{5.58}$$

Using $f = 8g/C^2$ results in:

$$\frac{1}{C^2} = \frac{1}{(C')^2} + \frac{1}{(C'')^2} \tag{5.59}$$

When using the same reasoning Manning's n results in:

$$n = n' + n''$$ (5.60)

5.2.9 *Determination of the friction factor*

A summary of the most widely used methods to predict the friction factor includes the methods:
– Van Rijn (1984c);
– Brownlie (1983);
– White, Paris and Bettes (1979);
– Engelund (1966).

The methods of Engelund, White et al. and Brownlie predict the friction factor as a function of the flow condition and sediment size. No explicit bed form characteristics are required. The van Rijn method is based on flow conditions and sediment size as well as on bed form and grain-related parameters such as bed form, length and height. More details about the methods that can be used to predict the friction factor are given in Appendix B.

The four methods to predict the flow resistance have been compared to find the most appropriate method for situations similar to those encountered in irrigation canals. The comparison has been based on field and flume data, and includes:
– RIJ (van Rijn, 1984c);
– BRO (Brownlie, 1983);
– WBP (White, Paris and Bettess, 1979);
– ENG (Engelund, 1966).

The accuracy of the prediction methods is evaluated by:

$$\frac{C_{measured}}{f} \leq C_{predicted} \leq C_{measured}{}^{*}f$$ (5.61)

$$Accuracy = \frac{\text{Number of well predicted values}}{\text{Total number of values}}$$ (5.62)

where:
$C_{measured}$ = measured de Chézy coefficients from the data set compiled by Brownlie (1981a);
$C_{predicted}$ = predicted de Chézy coefficients as determined by one of the four methods;
f = error factor

The overall accuracy of each prediction method is used to draw some conclusions concerning the applicability of each method:
• the four methods use only the bottom friction for the prediction of the friction factor. The B/y ratio of all the data used is larger than 10 and

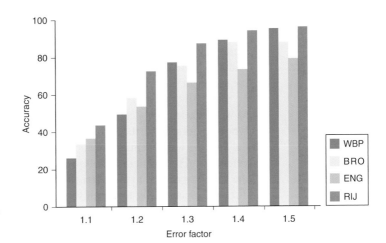

Figure 5.15. Accuracy of the methods to predict the friction factor for different error factors f.

the effect of the sidewalls is considered to be negligible. However, for a non-wide canal the sidewalls will have a significant effect on the friction factor. Especially the varying water depth above the sidewalls and the roughness of the sidewalls without any bed form will require a weighted friction factor;

• the van Rijn method (1984c) to predict the friction factor appears to give the best results when compared with the data from the Brownlie set. Approximately 41%, 71%, 88%, 97% and 98% of *well-predicted* bed forms for an error factor (f) of 1.1, 1.2, 1.3, 1.4 and 1.5 respectively are obtained (Figure 5.15). The van Rijn method shows good results over the whole range of measured friction factors.

5.3 GOVERNING EQUATIONS FOR SEDIMENT TRANSPORT

5.3.1 *Introduction*

Sediment transport predictors have often been applied on the basis of the mean velocity as a variable without considering the variation in channel geometry, distribution of the flow velocity and/or sediment transport in the cross section (Simons and Senturk, 1992). Sediment transport predictors have been developed for wide canals and they consider a channel with an infinite width without taking into account the effect of the sidewalls on the water flow and sediment transport. This ignorance of the effect of the side-walls implies that the velocity and the sediment transport are considered to be constant at any point of the cross section. Under these assumptions a uniformly distributed shear stress on the bottom and an identical velocity and sediment transport at any point over the canal width are assumed. In this way, these variables can be easily expressed per unit width. For other, non-wide canals, the shape will have an important impact on the water flow and sediment transport. The existence of sidewalls and the varying water depth on the sides will cause a non-uniform distribution over the width for both the shear stress and the velocity, and as a consequence, for the sediment transport.

Generally, the most common shape of an irrigation canal is trapezoidal. Most of these canals cannot be considered as a wide canal and for this type of cross section, the imposed boundary condition for the velocity and the varying water depth on the sides will affect the shear stress and the velocity and sediment distribution in the lateral direction. In other words, the variables related to the water flow and sediment transport vary in that direction, specifically for small values of the B/y ratio.

5.3.2 *Initiation of sediment movement*

Sediment can be transported in equilibrium or non-equilibrium conditions. Equilibrium conditions means that for specific hydraulic conditions the flow conveys a certain amount of sediment without deposition or erosion. Sediment transport predictors are supposed to describe the transport for the equilibrium conditions. The sediment transport under non-equilibrium conditions defines how the flow conveys a certain amount of sediment, as well as the erosion and deposition that take place at the same time.

The aim of sediment research is to predict the sediment transport in relation to a known sediment input. Three modes of sediment motion can be distinguished: *rolling and sliding, saltation and suspension*. The modes of motion are related to the flow conditions and the bed material, especially the hydrodynamic forces, which are expressed in terms of mean velocity or bottom shear stress acting on a bed. Firstly, the hydrodynamic forces have to reach a critical or threshold value for the initiation of motion, before a small increase of the forces will put the grain or aggregate into motion by irregular jumps or by rolling of the particles. Secondly, when the hydrodynamic forces reach a threshold value for the initiation of suspension then the sediment particles start to diffuse into the water flow. Based on the three modes of motion two different types of sediment transport can be defined:
– bed load;
– suspended load.

Bed-load transport

The transport of bed material particles by a water flow can be in the form of bed load or suspended load, depending on the size of the bed material particles and the flow conditions. In natural conditions there is no sharp division between the bed-load or suspended load transport.

Bagnold (1966) defines bed load as the particles which are in successive contacts with the bed and the processes are governed by gravity. The bed-load transport includes the *rolling and sliding and the saltation* modes in the bed layer. Bed-load transport processes and the formula to compute the yield have been given by many scientists. The simplest yet fairly reliable empirical equation is based on experimental data by Meyer-Peter (Raudkivi, 1990). Einstein (1950) introduced statistical methods to represent the turbulent behaviour of the flow and gave a bed-load function. Bagnold's (1966) equation is based on a physical concept and analysis. He introduced an energy concept and related the sediment transport rate to the work done by the fluid. Equations by Engelund (1966), Ackers and White (1973) and Yalin (1977) are mainly based on the Einstein or Bagnold concepts but deduced using dimensional analysis. Van Rijn (1984a) solved the equations of motion of an individual bed-load particle and computed the saltation characteristics and particle velocity as a function of the flow conditions and the particle diameter for plane bed conditions.

Suspended load transport

Suspended load is that part of the sediment transport that moves with the water flow without contact with the bottom. It includes the suspension mode and the wash load, which consist of cohesive and very fine sediments (smaller than 0.05 mm) and which tend to be suspended by Brownian motion (Raudkivi, 1990).

Transport of suspended sediment takes place when the bed shear velocity (u_*) exceeds the particle fall velocity (w_s). The particles can be lifted to a level at which the upward turbulent forces will be comparable to or higher than the submerged particle weight and as a consequence the contact of the particle with the bed is occasional and random in the suspension mode. The particle velocity in the longitudinal direction is almost equal to the fluid velocity. Usually, the behaviour of the suspended sediment particles is described in terms of sediment concentration, which is the solid volume per unit fluid volume or the solid mass (kg) per unit fluid volume. The principle feature that distinguishes the suspended sediment transport from the bed-load transport is the time taken for the suspension to adapt to changes in flow conditions (Galappatti, 1983).

According to Bagnold (1966), suspension will occur for a bed shear velocity (u_*) equal or larger than the particle fall velocity (w_s), while van Rijn (1984b) argues that suspension will start at considerably smaller bed shear velocities. From his experimental data van Rijn (1984b) proposed the following conditions for the initiation of suspension:

$$\frac{u_{*,cr}}{w_s} = 4D_*^{-1} \quad \text{for } 1 < D_* \leq 10 \tag{5.63}$$

$$\frac{u_{*,cr}}{w_s} = 0.4 \qquad \text{for } D_* > 10 \tag{5.64}$$

Discharge of the sediment per unit width can be written as:

$$q_{s,c} = \int_a^h uc\,dz \tag{5.65}$$

$$q_{s,c} = c_a\bar{u} \int_a^h \frac{u}{\bar{u}} \frac{c}{c_a} dz \tag{5.66}$$

where:
  c = sediment concentration at height z from the bed
  c_a = reference concentration at reference height a above the bed level
  h = water depth (m)
  $q_{s,c}$ = volumetric suspended transport (m^2/s)
  u = velocity at height z above the bed (m/s)
  $\bar{u}$ = depth-averaged flow velocity (m/s)

For the solution of the equation, the velocity profile, concentration profile and concentration at reference level should be known. For steady and uniform conditions of water and sediment, the concentration profile of sediment in the vertical is in equilibrium. A number of relations are available for the prediction of suspended sediment transport rates. Some of them are based on analytical approaches, but they still need experimental results to derive certain parameters.

Initiation of motion

Due to the fact that the sediments entering irrigation canals are from external sources (for instance, rivers), the particle size of the sediment is usually different from the parent bed material. The incoming particle size depends on the operation of the sediment trap or intake structure at the head of the canal network. Normally the sediments entering into the irrigation canals are in the range of fine sand, silt and clay (Worapansopak, 1992). Larger particles are preferably excluded from entering the canal system by a careful skimming of the water at the intake or have been allowed to settle in a sediment trap in the first reach of the canal system (Dahmen, 1994). In view of these provisions at the head of an irrigation network, the sediment sizes are assumed to be in the range of 0.05 mm $< d_{50} < 0.5$ mm. It is also assumed that only non-cohesive material will be present in the irrigation system, despite some degree of cohesion that is present for the smaller particle sizes.

Yalin (1977) expressed this process by:

$$\text{No motion} \qquad \tau < \tau_{\text{cr}} \tag{5.67}$$

$$\text{Bed load transport} \qquad \tau_{\text{cr}} \leq \tau \leq \tau'_{\text{cr}} \tag{5.68}$$

$$\text{Bed and suspended load transport} \qquad \tau \geq \tau'_{\text{cr}} \tag{5.69}$$

where:
$\tau = $ bottom shear stress
$\tau_{\text{cr}} = $ critical shear stress for initiation of motion
$\tau'_{\text{cr}} = $ critical shear stress for initiation of suspension.

In reality, there is not *one* critical value at which the motion and sus-pension suddenly begins, but it fluctuates around an average value. The movement of the particles is highly unsteady and depends on the turbu-lence of the flow. It is not possible to give a single value that presents zero movement and for that reason it is easier to define the condition for initiation of motion as the one below a certain value for which the sediment transport rate has no practical meaning (Paintal, 1971).

Several authors have developed theories to explain the initiation of motion. Most of these are based either on a critical depth-averaged velocity or on a critical bed shear stress. The theories based on the critical velocity require water depths that completely satisfy the flow condition at which the initiation of motion occurs, whereas the theories based on critical shear stress describe the flow condition for the initiation of motion by using a single critical value for the shear stress. ASCE (1966) recommends that data on critical shear stress should be used wherever possible. Among the theories based on critical shear stress, the Shields' diagram is the most widely accepted criterion to describe the conditions for initiation of motion of uniform and non-cohesive sediment on a horizontal bed.

The Shields' diagram (see Figure 5.16) gives the relation between the critical mobility Shields' parameter (θ_{cr}) and the dimensionless particle Reynold's number (Re$_*$). The particle Reynold's number represents the hydraulic condition on the bed and is based on the grain size and the shear velocity. The initiation of motion will occur when the mobility Shields' parameter (θ) is greater than the critical mobility Shields' parameter (θ_{cr}).

These parameters are expressed by:

$$\theta_{\text{cr}} = \frac{u_{*\text{cr}}^2}{(s-1)gd_{50}} = \frac{\tau_{\text{cr}}}{(s-1)\rho g d_{50}} \tag{5.70}$$

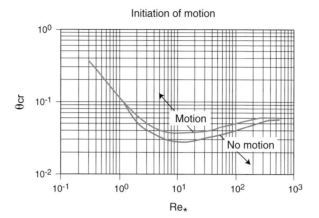

Figure 5.16. Shields' diagram for initiation of motion (after van Rijn, 1993).

$$\mathrm{Re}_* = \frac{u_{*cr}d_{50}}{v} \tag{5.71}$$

$$\theta = \frac{u_*^2}{(s-1)gd_{50}} = \frac{\tau}{(s-1)\rho gd_{50}} \tag{5.72}$$

$$u_* = \sqrt{g\,hS_{\mathrm{o}}} \tag{5.73}$$

The use of the Shields' diagram is not very practical, since the u_* value appears in both parameters and in both axes of the diagram, and can only be solved by trial and error. This imperfection of the Shields' diagram is eliminated by introducing the particle parameter D_* which is represented by (Yalin, 1977):

$$D_* = \frac{\mathrm{Re}_*^2}{\theta} = \left[\frac{(s-1)g}{v^2}\right]^{1/3} d_{50} \tag{5.74}$$

In this way the critical mobility parameter θ_{cr} will be expressed as a function of D_*. Once D_* is known the Van Rijn (1993) relationships can be used to determine the value of Θ_{cr} and to solve the equation for τ_{cr}. According to van Rijn the relationships between θ_{cr} and D_* are:

$$\theta_{\mathrm{cr}} = 0.24D_*^{-1} \quad \text{for } 1 < D_* \leq 4 \tag{5.75}$$

$$\theta_{\mathrm{cr}} = 0.14D_*^{-0.64} \quad \text{for } 4 < D_* \leq 10 \tag{5.76}$$

$$\theta_{\mathrm{cr}} = 0.04D_*^{-0.1} \quad \text{for } 10 \leq D_* \leq 20 \tag{5.77}$$

$$\theta_{\mathrm{cr}} = 0.013D_*^{0.29} \quad \text{for } 20 \leq D_* \leq 150 \tag{5.78}$$

$$\theta_{\mathrm{cr}} = 0.055 \quad \text{for } D_* > 20 \tag{5.79}$$

Initiation of suspension: van Rijn (1984b) describes the initiation of suspension by:

$$\theta'_{cr} = \frac{16}{D_*^2} \frac{w_s^2}{(s-1)gd_{50}} \quad \text{for } 1 < D_* \leq 10 \tag{5.80}$$

$$\theta'_{cr} = 16 \frac{w_s^2}{(s-1)gd_{50}} \quad \text{for } D_* > 10 \tag{5.81}$$

These equations can also be written as:

$$\frac{u_{*,cr}}{w_s} = 4D_*^{-1} \quad \text{for } 1 < D_* \leq 10 \tag{5.82}$$

$$\frac{u_{*,cr}}{w_s} = 0.4 \quad \text{for } D_* > 10 \tag{5.83}$$

where:
   θ = mobility Shields' parameter (dimensionless)
   θ_{cr} = critical mobility Shields' parameter (dimensionless)
   Re_* = particle Reynolds number (dimensionless)
   s = relative density (ρ_s/ρ)
   u_* = local shear velocity (m/s)
   w_s = fall velocity (m/s)
   D_* = particle parameter (dimensionless)
   d_{50} = median diameter (m)
   g = acceleration due to gravity (m/s^2).

Irrigation canals are manmade canals and their design takes into account aspects related to the irrigation criteria and sediment transport. On the one hand, the canals should meet the irrigation requirements and on the other hand, no deposition of the sediment entering into the system or scouring of the parent material should occur. Suggested values for non-scouring bottom shear stress are available in the literature. Kinori (1970) and Chow (1983) give minimum values for the non-scouring shear stress for water containing fine sediments in the range between 1.5 N/m^2 (fine sand, sandy loam) and 15 N/m^2 (hard clay and gravel). Dahmen (1994) suggests a maximum value for the design of irrigation canals of 3–4 N/m^2. Even though the design values for the shear stress can be reduced due to changes in the operation strategies during the irrigation season, the value of the remaining shear stress is still high enough to initiate motion and further suspension of the previously deposited sediment. However, it is no longer so high as to produce scouring of the parent material of the canal.

Figure 5.17 shows the Shields' curve for initiation of motion and the criteria used by van Rijn (1993) to initiate suspension. This figure presents a range of shear stresses between 1 N/m² and 4 N/m², which is a range of shear stress commonly used for the design of irrigation canals. In addition it clearly shows that typical flow conditions in irrigation canals are large enough to produce suspension of sediment particles. The sediment in irrigation canals is transported in two modes: suspended load and bed load.

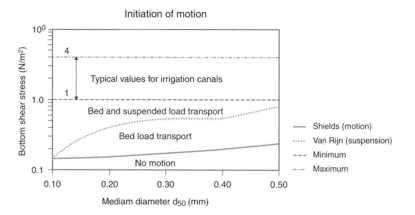

Figure 5.17. Initiation of motion and suspension and values of shear stress as function of D_*.

In irrigation canals the sediment is transported both as bed load and as suspended load. Therefore, any predictor to estimate the sediment transport should be able to compute either implicitly the total transport (bed load + suspended load) or the bed load and suspended load separately. Only for very fine sediment ($d_{50} < 0.1$ mm) can a suspended sediment transport predictor be used to estimate the sediment transport capacity of irrigation canals.

5.3.3 *Sediment transport in non-wide canals*

In general the reliability of sediment transport predictors is low and at best they can provide only rough estimates. A probable error in the range of 50–100% can be expected even under the most favourable circumstances (Vanoni, 1975). The error is expected to increase further if the calculations are based upon average values of flow and sediment parameters. Several assessments of sediment transport formulas have been made (Brownlie, 1981b; Yang and Molinas, 1982; Van Rijn, 1984b; Yang and Wan, 1991) and each provide different results. Woo and Yu (2001) compared the results assessed by different researchers and found that there is no universally accepted formula for the prediction of sediment transport. Most of them are based upon laboratory data of limited sediment and water flow ranges. Hence they should be adjusted to make them compatible for the specific purposes; otherwise the predicted result will be unrealistic.

The distribution of shear stress along the boundary is not constant, in contrast to the general assumptions made in the calculations. Even in flumes for sediment transport experiments the boundary shear is not constant due to the presence of bed forms. For laboratory data a sidewall correction is applied, so that the data can be treated as that of a wide canal and accordingly empirical relationships are developed. Different techniques are in use for the sidewall correction (Einstein, 1942; Vanoni and Brooks, 1957), where the main objective is to find an average bed shear stress after making due allowance for the friction of the sidewall.

Engelund and Hansen (1967) pointed out that even after the theoretical side wall correction, the experimental flume data should be considered with caution, since the shear stress not only depends on the relative roughness of bed and walls but also on the B/y ratio of the flume. Moreover Brownlie (1981a) mentions that field data have slightly higher sediment concentrations than laboratory data for similar ranges of dimensional groups. For a theoretical analysis of the discrepancy in laboratory and field observations, he used a typical river section and showed that the difference was due to the changing water depth along the wetted perimeter compared to a constant depth in a laboratory flume.

An analysis of the applicability of the sediment transport predictors in irrigation canals should consider two aspects; namely the side slope m and the B/y ratio. The majority of the canals have a trapezoidal cross section with side slopes ranging from 1:1 to 1:4 depending upon the soil type and bank stability with the exception of small and lined canals that may be rectangular. The changing water depth on the sides will influence the overall shear stress distribution along the perimeter. This effect is more pronounced if the B/y ratio is small. The majority of the irrigation canals are non-wide and their B/y ratio is less than 8 (Dahmen, 1994). Hence, the assumption of a uniform velocity and sediment transport across the section and expressing them per unit width of canal is not correct.

For the same hydraulic and sediment characteristics different sediment transport predictors give widely varying results. For one specific condition one predictor gives better results than the other and it is not possible to adjust all the predictors to produce the same value for a given condition. Hence, the purpose of the adjustment is to adapt the equation for a specific canal condition and with this fine-tuning the predictability should be improved.

5.3.4 *Effects of the canal geometry and flow characteristics on the sediment transport*

For the flow conditions and sediment characteristics prevailing in the irrigation canals, Méndez (1998), after evaluating the available total load predictors with field and laboratory data, concludes that the predictors given by Brownlie, Ackers and White, and Engelund and Hansen are better compared to other predictors. However, a prediction within an error factor of less than 2 was not possible.

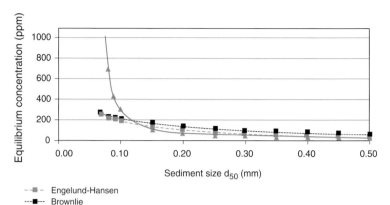

Figure 5.18. Effect of sediment size (d_{50}) on the prediction of the equilibrium concentration.

The effect of the canal geometry and flow characteristics on the sediment transport has been analysed for the sediment size, velocity, B/y ration and side slope (Paudel et al., 2013). The sediment size normally found in an irrigation canal is 0.05 mm (50 μm) to 0.5 mm. For the evaluation of the effect of the sediment size on the predictability of the three above mentioned sediment transport predictors the discharge, bed width, bed slope, side slope and the roughness of a canal was kept constant and the equilibrium concentration was computed for the three total load predictors (see Figure 5.18). The Brownlie predictor shows an almost linear variation for sediment sizes between 0.10 and 0.50 mm. Ackers and White predict high values for sediment sizes below 0.15 mm. The predictability of the three predictors for sediment sizes greater than 0.15 mm is comparable.

The sediment transport in relation to the change in velocity was evaluated for the same discharge, bed width and side slope, while the bottom slope was increased to increase the velocity. The Froude number in all cases was smaller than 0.5. The results are presented in Figure 5.19.

The influence of the velocity on the prediction of the transport capacity is for the Brownlie and Engelund-Hansen predictors almost the same, while the Ackers-White predictor is more sensitive to changes in the velocity. The change is more significant for higher velocities (greater than 0.70 m/s), which correspond to a Froude number of 0.24. The Froude number in irrigation canals normally varies from 0.2 to 0.5.

To appraise the effect of a change in the B/y ratio and side slope m on the sediment transport, the discharge, bed slope and the roughness were kept constant. For each side slope, the bed width was changed to obtain different B/y ratios. A change in bed width also changed the flow velocity. The results for three different side slopes of 1:1, 1:1.5 and 1:2 are presented in Figure 5.20. The Engelund-Hansen and Ackers-White methods show similar prediction trends. For a small B/y ratio the transport capacity is

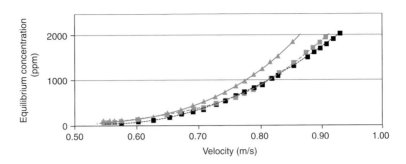

Figure 5.19. Effect of velocity in the prediction of the equilibrium concentration.

also small and the capacity will increase with the B/y ratio. For B/y ratios larger than about 5 the transport capacity remains almost constant for the Engelund-Hansen method and it decreases for the Ackers-White method. The Brownlie predictor shows a larger transport capacity for the smallest B/y ratio and it decreases with an increasing B/y ratio.

All three methods predict that more gentle side slopes give lower transport capacities. For a given B/y ratio steeper side slopes result in an increase in water depth and flow velocity and hence, in the transport capacity.

5.3.5 *Velocity distribution in a trapezoidal canal*

The velocity distribution in a trapezoidal canal is not only influenced by the effect of the boundary shear stress, but also by the changing water depth along the slope. This influence is significant for non-wide canals and the concept of a uniform velocity distribution across the section cannot be used. Moreover the roughness is not constant along the perimeter due to the presence of bed forms and protection works or vegetation on the sides. Einstein (1942) suggested that the total area of these canals can be divided into an area that corresponds to the bed (A_b) and the other to the sidewalls (A_w). The average shear stress in the bed and on the sidewalls for a constant friction slope can be written as:

$$\tau_b = \rho g R_b S \tag{5.84}$$

$$\tau_w = \rho g R_w S \tag{5.85}$$

The shear stress depends on the distance R, which means that the surplus energy in any volume of flowing water will be dissipated by the shortest-distance boundary.

Based on this subjective concept, the surplus energy at any point in the column CD above the line EF will be dissipated by the side slope and

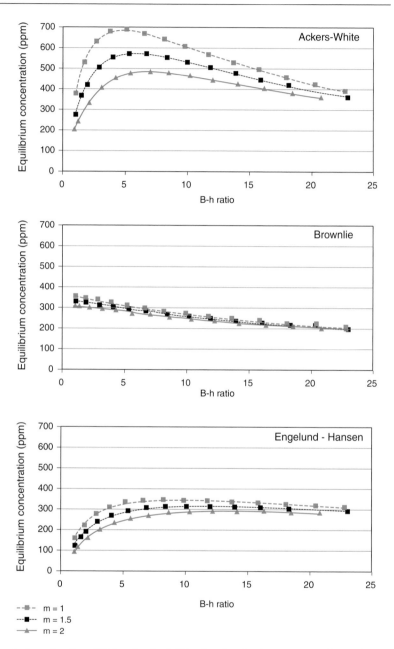

Figure 5.20. Effect of *B*/*y* ratio and side slope on the prediction of equilibrium concentration.

below the line EF by the bed. The local velocity at any point A in the column CD (Figure 5.21) is given by Yang and Lim, 1997:

$$\frac{u_{(z',y')}}{u_*} = 2.5 \ln\left(f \frac{z'}{z_0} \right) \tag{5.86}$$

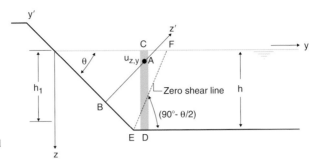

Figure 5.21. Definition sketch for calculation of depth-averaged velocity (Yang et al., 2004).

$$f = \frac{u_{*y}}{u_{*h}} \tag{5.87}$$

where:
$u_{z',y'}$ = velocity at points in the shaded column
$\quad u_*$ = overall mean shear velocity (m/s)
$\quad u_{*y}$ = local shear velocity based on local boundary shear stress (m/s)
$\quad u_{*h}$ = local shear velocity at the centre of the canal (m/s)
$\quad z_0 = k/30$ for a rough boundary, where k is the roughness height (m)

To compute the depth-averaged velocity over the column CD, the following assumptions are made:
- the roughness in the bed and on the sides are equal, so the line of zero shear (EF) is the bisector of the angle between the bed and side slope;
- the local shear velocity can be replaced by the local average shear on either the bed or the sidewall.

Integration of the equation along the column CD gives the relation for the depth-averaged velocity (u_y) in a stream column (Yang et al., 2004):

$$\frac{\overline{u}_y}{u_*} = 2.5 \ln\left(f\frac{h_1}{z_0}\cos\theta\right) - 2.5(1+\beta) \tag{5.88}$$

In non-wide canals and near the sidewalls in wide canals the maximum velocity is located below the free surface, which is known as the dip phenomena. The second term in the right hand side of the equation accounts for the dip phenomena by β, being the correction factor for the dip phenomenon. For a smooth canal the value of β is given by:

$$\beta = \frac{1.3}{\sin\theta}\exp\left(-\frac{y}{h_1}\right) \tag{5.89}$$

The velocity distribution in a trapezoidal, earthen irrigation canal has been measured (Paudel, 2010) and for the rough canal the coefficient β has been determined:

$$\beta = \frac{0.8}{\sin \theta} \exp\left(-\frac{y}{h_1}\right)$$

(5.90)

where:

h_1 = water depth at point 1 (m)
θ = angle made by side slope with the water surface
β = correction factor for the dip phenomenon

Figure 5.22 shows the velocity distribution as predicted by the equation and measured in the irrigation canal.

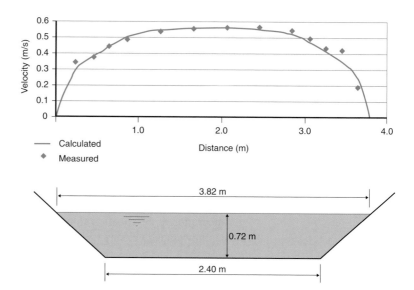

Figure 5.22. Examples of calculated and measured velocity distribution in an earthen irrigation canal.

5.3.6 *Exponent of the velocity in the sediment transport predictors*

Sediment transport predictors are in different forms and complexities depending upon the assumptions and the basic approaches used in the derivation of the predictor. There is no general agreement on the type of variables that are required to define the sediment transport, but the most frequently used ones are:

$$q_s = f(u, h, S, \rho, v, \rho_s, d_{50}, g, \sigma_g)$$

(5.91)

The sediment transport per unit width can be approximated by the power law, where the coefficients M and N are supposed to be locally constant (de Vries, 1987):

$$q_s = MV^N \tag{5.92}$$

Differentiation of the sediment transport equation with respect to V results in:

$$\frac{dq_s}{dV} = MNV^{N-1} \tag{5.93}$$

This gives:

$$N = \frac{dq_s}{dV}\frac{V}{q_s} \tag{5.94}$$

where:
q_s = sediment transport capacity per unit width (m³/s·m)
V = mean velocity (m/s)
M, N = coefficient/exponent depending on the water flow and sediment characteristics.

For some predictors the value of N can be directly derived from the equation.

For Engelund and Hansen (1967) N follows from:

$$q_s = \frac{0.05V^5}{(s-1)^2 g^{0.5} d_{50} C^3} \tag{5.95}$$

which gives $N = 5$.

According to Klaassen (1995), the equation for N can be used for more complex sediment predictors such as Ackers and White (1973), van Rijn (1984a, 1984b) and Brownlie (1981b).

The Ackers and White predictor given by (see Annex A) is:

$$q_s = G_{gr} V d_{35} \left(\frac{V}{u^*}\right)^n \tag{5.96}$$

Differentiating and comparing the results with the equation for N gives the following relation for N (de Vries, 1985):

$$N = 1 + \frac{m' F_{gr}}{(F_{gr} - A)} \tag{5.97}$$

The dimensionless mobility parameter F_{gr} is given by:

$$F_{gr} = \frac{u_*^n}{\sqrt{gd_{35}(s-1)}} \left[\frac{V}{\sqrt{32}\log\left(\frac{10h}{d_{35}}\right)} \right]^{1-n} \qquad (5.98)$$

F_g, is the grain Froude number and is given by:

$$F_g = \frac{V}{[(s-1)gd_{50}]^{0.5}} \qquad (5.99)$$

The critical grain Froude number F_{gcr} is:

$$F_{gr} = 4.596\tau_{*0}^{0.5293}S^{-0.1405}\sigma_s^{-0.1696} \qquad (5.100)$$

Similarly the Brownlie's predictor is given by (see Appendix A):

$$q_s = \frac{0.007115q}{s}c_f(F_g - F_{gcr})^{1.978}S^{0.6601}\left(\frac{R}{d_{50}}\right)^{-0.3301} \qquad (5.101)$$

Differentiation gives:

$$N = 1 + \frac{1.978F_g}{F_g - F_{gcr}} \qquad (5.102)$$

The assessment of N (Klaassen, 1995; Méndez, 1998) shows that the value N depends upon the flow conditions and sediment characteristics. In most cases, except for Engelund-Hansen (Bagnold, 1966) N increases with a decrease in flow velocity and in particle size (see Figure 5.23). For a higher velocity the value of N remains fairly constant for a specified particle size.

For the Brownlie predictor the exponent N is for a geometric standard deviation for the grain Froude numbers F_g larger than 10, independent of the bed slope and sediment size (see Figure 5.23). A standard deviation for F_g more than 10 refers to a velocity that is slightly larger than that required for the initiation of motion. Moreover the value of N is always more than 3 for sediment smaller than 0.50 mm and Froude numbers smaller than 0.6.

The exponent N for the Ackers-White predictor is more sensitive to the sediment size (Figure 5.24). Near the initiation of motion the value of N cannot be determined. N is always more than 4 for Froude numbers less than 0.6, which is also the upper limit of the normal flow conditions in irrigation canals.

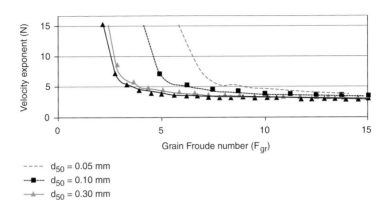

Figure 5.23. Variation of N
with grain Froude number for
Brownlie's predictor.

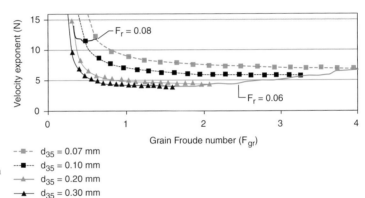

Figure 5.24. Variation of N with
grain Froude number for the
Ackers-White predictor.

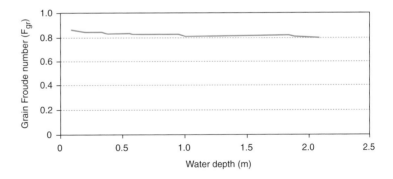

Figure 5.25. Variation of the
grain Froude number F_{gr} with
water depth (Paudel, 2010).

Moreover, the parameter F_{gr} in the Ackers-White method also depends
on the water depth and in a trapezoidal cross-section the velocity changes
together with the water depth on the side slopes. However, calculations
(Figure 5.25) show that the change in water depth has a very small effect
on the parameter F_{gr} and hence can be neglected.

5.3.7 Correction factor for the prediction of the total sediment transport

The cross-section integrated approach is based on the assumption of a quasi-two-dimensional model and the cross section is composed by a series of parallel stream tubes (see Figure 5.26). The mean velocity over the whole cross section will be replaced by the local, non-uniform velocity (u_i). Within each stream tube, the lateral velocity distribution is considered to be uniform and therefore can be described in a one-dimensional way. The sediment transport in each stream tube is considered as a function of the flow in that stream tube only without taking into account any diffusion in the y-direction. The sediment transport per unit width given by $q_s = MV^N$ will be replaced by each local velocity $q_s = Mu_i^N$. The difference between the total transport values from the mean velocity and the local velocities requires a correction factor α in order to equal both values for the total sediment transport. The division of the entire cross section into small columns of finite width dy (Figure 5.26) results in the sediment transport through that stream column and the velocity u that is averaged over the local depth.

$$q_s = Mu^N \tag{5.103}$$

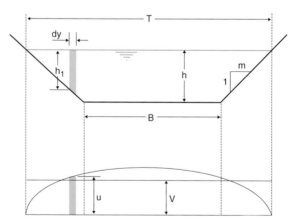

Figure 5.26. Schematized stream column and velocity distribution in a non-wide canal (Paudel, 2010).

Summation of the sediment transport of all the stream tubes will give the total sediment transport of the cross section.

$$Q_s = \int_T Mu^N \, dy \tag{5.104}$$

Using the average of the variables over the whole cross section will also result in the total sediment transport through the whole section:

$$Q_s = B_s MV^N \tag{5.105}$$

where:

B_s = the sediment transport width of the section (m)

T = the top width of the canal (m)

dy = the width of the stream tube (m)

V = mean velocity (m/s)

Q_s = sediment transport capacity for the whole cross section (m³/s)

q_s = sediment discharge of the stream tube i per unit width (m³/s)

The second sediment transport (5.105) is not equal to the transport given by the aforementioned equation (5.104) due to the nonlinear relationship between the sediment transport and the velocity. Introduction of a correction factor α is necessary to equalize the total transport computed by the two methods:

$$\alpha_s = \frac{\int_T Mu^N \, dy}{B_s MV^N} \tag{5.106}$$

This correction factor α for the prediction of the total sediment transport is a function of the velocity distribution in the cross section and the exponent N in the sediment transport predictor.

The factor α will take into account the influence of the non-uniform velocity distribution in trapezoidal, non-wide canals. In rectangular canals the influence of the side slope on the velocity distribution will be small, but the non-uniform velocity distribution will still exist especially for a small B/y ratio.

Use of these simplified equations is correct to understand the effect of the cross section on the sediment transport, but for a final adjustment they should be analysed separately. The coefficient M and exponent N have different values that depend upon the variables in the specific sediment predictors. Hence, the correction factor α for changes in side slope m, B/y ratio, sediment size and velocity exponent (N) should be evaluated separately for each predictor. The evaluation of the correction factor for the total sediment transport capacity will be given for the Ackers-White, Brownlie, and Engelund-Hansen methods (Paudel, 2010).

The following range of hydraulic and sediment characteristics will be used:

o Froude number 0.05 to 0.5;

o sediment size (d_{50}) 0.075 to 0.5 mm;

o bed width to water depth ratio $B/y = 2$ to 12;

o side slope $m = 0$ to 3;

o de Chézy roughness coefficient is 35 to 60;

o number of stream tubes 40;

o geometric standard deviation of the bed particle size σ_g is 1.2 to 1.8.

The factor α will be determined for each of the three sediment transport predictors.

- The exponent N in the Engelund-Hansen predictor is constant and is 5. The correction for this predictor is a function of the B/y ratio and side slope m. The correction is for:

trapezoidal canal: $\alpha_s = 1.2785(B/h)^{-0.0937}m^{0.078}$ (5.107)

rectangular canal: $\alpha_s = 1.2(B/h)^{-0.0663}$ (5.108)

- In the Ackers-White predictor the correction factor α is a function of the B/y ratio, velocity exponent N, side slope m and sediment size d_{50}. For the analysis the sediment size d_{50} is replaced by the dimensionless grain parameter D and the exponent N varies from 3 to 10. All the other hydraulic and sediment characteristics are the same as for the Engelund-Hansen predictor. Using a non-linear regression analysis the correction factor α can be given by:

trapezoidal canal: $\alpha_s = 0.396(B/h)^{-0.1012}N^{0.7514}m^{0.0541}(\log D_*)^{0.2427}$

(5.109)

rectangular canal: $\alpha_s = 0.0868(B/h)^{-0.1699}N^{1.3175}D_*^{0.3153}$ (5.110)

- The correction factor for the Brownlie predictor is a function of the B/y ratio, velocity exponent N and side slope m. The exponent N varies from 3 to 6. All the other hydraulic and sediment characteristics are the same as for the Engelund-Hansen predictor. A non-linear regression analysis gives the following correction factor α for:

trapezoidal canal: $\alpha_s = 1.023(B/h)^{-0.0898}N^{0.1569}m^{0.078}$ (5.111)

rectangular canal: $\alpha_s = 0.8492(B/h)^{-0.0361}N^{0.2106}$ (5.112)

5.3.8 *Predictability of the predictors with the correction factor*

Literature presents several procedures that have been developed to compute the total sediment discharge in open channels and here they will be named procedures 1, 2 and 3.

o *Procedure 1*: the sediment discharge per unit width (q_s) is calculated by using the hydraulic radius R as the characteristic variable for the water flow. The average width represents the canal width and the total sediment transport is determined by the multiplication of these variables:

$Q_s = q_s^* B_{av}$ with $q_s = f(R)$ (5.113)

o *Procedure 2*: the sediment transport per unit width (q_s) is calculated by using the water depth as the characteristic variable. Next, the total sediment transport Q_s is calculated by multiplying the sediment transport per unit width (q_s) and the bottom width B:

$Q_s = q_s^* B_{av}$ with $q_s = f(h)$ (5.114)

o *Procedure 3*: the total sediment transport is determined by computing the sediment transport capacity of each stream tube and added up for the whole area (the cross-section integrated approach):

$$Q_s = \int_{i=1}^{n} q_{s(i)} dy \qquad (5.115)$$

The total sediment transport for the total cross section is found by:

$$Q_s = \alpha B q_s \quad \text{with } q_s = f(v) \qquad (5.116)$$

where:
$\alpha =$ correction factor for calculating the total sediment transport in a non-wide canal.
$B =$ bottom width (m)
$v =$ mean velocity (m/s)
$Q_s =$ sediment transport capacity for the whole cross section (m³/s)
$q_s =$ sediment transport capacity per unit width (m³/s · m)

In this method the sediment transport follows from the quantity obtained from the predictors by assuming that the canal is wide and then the transport is corrected for the non-wide conditions and the side slope.

In order to compare the three procedures in terms of their ability to compute the total sediment transport in non-wide canals, these procedures have been applied to a selected set of laboratory data. A correct method to compute the sediment transport should take into account the effect of the cross section on the velocity distribution and the non-linear relationship between the velocity and the sediment transport. Paudel (2010) selected specific data from the Brownlie (1981a) compilation with flume and field data. The selection criteria are based on the flow conditions and sediment characteristics that normally exist in irrigation canals and include:

- sediment size less than 0.5 mm;
- Froude number less than 0.5;
- B/y ratio less than 8;
- sediment concentration 100–1500 ppm;
- geometric standard deviation of the bed particle size σ_g less than 1.5;
- type of bed form is dune or less;
- data that include all the information needed for the sediment transport computation.

Figure 5.27 shows the sediment size and the range of concentrations in the data.

A total of 149 data sets were selected out of the Brownlie compilation (Table 5.8). All the data are for rectangular flumes with a maximum width of 2.4 m. The width for the sediment transport calculation in the three procedures given above will be equal to the bed width.

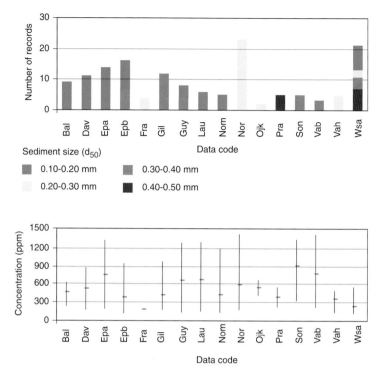

Figure 5.27. Characteristics of the selected data (Paudel, 2010).

Table 5.8. Selected data set for the evaluation of the sediment transport capacity computation procedures.

Investigator and year	Data code	No. of records
Gilbert, G.K. (1914)	GIL	12
U.S. Waterway Experiment Station (1935A)	WSA	35
U.S. Waterway Experiment Station (1936B)	WSS	17
Barton, J.R. and Lin, P.N. (1955)	BAL	9
Nomicos, G. (1957)	NOM	5
Vanoni, V.A. and Brooks, N.H. (1957)	VAB	4
Laursen, E.M. (1958)	LAU	6
Vanoni, V.A. and Hwang (1965)	VAH	6
Guy, H.P. et al. (1966)	GUY	8
Government of Pakistan (1966–69)	EPB	20
East Pakistan Water and Power (1967)	EPA	21
Franco, J.J. (1968)	FRA	7
Pratt, C.J. (1970)	PRA	8
Davies, T.R. (1971)	DAV	13
Onishi, Jain and Kennedy (1972)	OJK	4
Nordin, C.F. (1976)	NOR	26
Sony, J.P. (1980)	SON	5

In the comparison, the Ackers-White, Brownlie and Engelund-Hansen methods have been applied to predict the total sediment transport under equilibrium conditions. The three procedures have been compared on a relative basis. The sediment transport from the Brownlie data set is compared with the predicted sediment transport calculated by one of these three procedures. Then the predictability of each procedure is evaluated on the basis of well-predicted values within a certain accuracy range. For a given error factor (f) the upper and lower range are given by:

$$\frac{\text{measured value}}{f} \leq \text{predicted value} \leq \text{measured value} * f \qquad (5.117)$$

$$\text{accuracy} = \frac{\text{number of well predicted values}}{\text{total values}} \qquad (5.118)$$

The comparison (Figure 5.28) shows that the reliability of the predictors will improve when a correction factor is applied to incorporate the effect

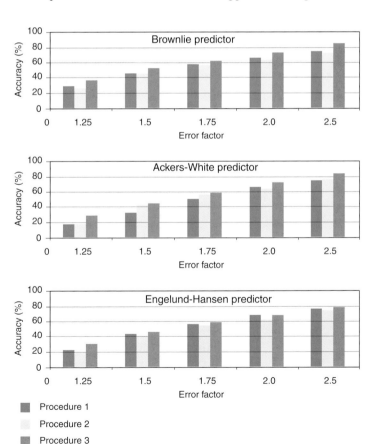

Figure 5.28. Evaluation of the three sediment transport predictors (Paudel, 2010).

of the B/y ratio (procedure 3). The predictability of the Brownlie predictor improves when the hydraulic mean depth R is used instead of the water depth. The accuracy of the Ackers-White predictor improves when the water depth is used instead of the hydraulic mean depth (R). The Engelund-Hansen prediction shows no significant change in accuracy for the two cases (procedures 1 and 2). The use of the bed width or the average width $(B + B_s)/2$ as the representative width will not influence the results as all the collected data are from rectangular cross sections. However, in a trapezoidal canal section the method of taking a representative width should have a significant effect on the total sediment transport. The most logical option could be the use of the average of the bed width and the top width.

As discussed earlier, sediment movement is a function of the hydraulic and sediment characteristics. A proper estimate of the design discharge and the roughness will support an accurate prediction of the hydraulic conditions. An understanding of the influence of the canal geometry on the sediment transport capacity for a given discharge will assist in the proper selection of the bed width, bed slope and side slope during the design.

Although the accuracy of the prediction will improve when the correction factor α is used for the sediment transport predictors, the accuracy of prediction will still be low. The accuracy is less than 50% for an error factor of 1.5. Maybe one predictor method predicts well for one data set, but for another data set the predictability is still very poor. This clearly indicates the limitation of the available sediment predictors. The predictors are derived from a limited range of hydraulic and sediment characteristics and their use outside this range should be made with utmost care. Therefore the predictors should be tested for the field conditions for which they are to be used and adjustments should be made once these modifications are found necessary.

5.3.9 *Sediment transport in non-equilibrium conditions*

Sediment transport in a steady and uniform flow has a unique equilibrium transport rate, but the transport rate under variable conditions differs from this steady uniform and equilibrium situation. Non-equilibrium conditions are very significant for the suspended load transport because the travel length of a suspended particle is in general much longer that the step length of a bed load motion (Nagakawa et al., 1989). While it is possible to relate the bed load transport to the local and instantaneous characteristics of water flow and bottom composition, no distinct relation can be established for the suspended load transport since this part of the transport mode is substantially influenced by upstream conditions (Armanini and Di Silvio, 1988).

Mathematical models for the simulation of non-equilibrium, suspended sediment transport in open canals might be based on the solution of:

- the 1-D, 2-D or 3-D convection-diffusion equation;
- depth-integrated models.

Depth-integrated models are based on a depth-integrated approach for the suspended sediment transport; the model describes how the mean concentration adapts in time and space towards the local mean equilibrium concentration. The suspended sediment transport model can be used together with the depth-averaged hydrodynamic equations.

Galappatti (1983) developed a depth-integrated suspended sediment transport model for suspended sediment transport in unsteady and non-uniform flow based on an asymptotic solution for the two-dimensional convection-diffusion equation in the vertical plane. In this model the vertical dimensions are eliminated by means of an asymptotic solution in which the concentration $c(x, z, \text{and } t)$ is expressed in terms of the depth-averaged concentration $c(x, t)$. The latter concentration is represented by a series of previously determined profile functions.

Among the depth-integrated models for suspended sediment transport this model has two advantages over others; firstly no empirical relation has been used during the derivation of the model and secondly all possible bed boundary conditions can be used (Wang and Ribberink, 1986). Moreover it includes the boundary condition near the bed, and hence an empirical relation for deposition/pick-up rate near the bed is not necessary (Ribberink, 1986). The partial differential equation that governs the transport of suspended sediment by convection and turbulent diffusion under gravity is given by (Galappatti, 1983):

$$\frac{\partial c}{\partial t} + u\frac{\partial c}{\partial x} + v\frac{\partial c}{\partial y} + w\frac{\partial c}{\partial z} = w_s\frac{\partial c}{\partial z} + \frac{\partial}{\partial x}\left(\varepsilon_x\frac{\partial c}{\partial x}\right) + \frac{\partial}{\partial y}\left(\varepsilon_y\frac{\partial c}{\partial y}\right)$$
$$+ \frac{\partial}{\partial z}\left(\varepsilon_z\frac{\partial c}{\partial z}\right) \tag{5.119}$$

Neglecting the diffusion terms other than the vertical, the equation for a two dimensional flow in the vertical plane becomes:

$$\frac{\partial c}{\partial t} + u\frac{\partial c}{\partial x} + w\frac{\partial c}{\partial z} = w_s\frac{\partial c}{\partial z} + \frac{\partial}{\partial z}\left(\varepsilon_z\frac{\partial c}{\partial z}\right) \tag{5.120}$$

Galappatti (1983) assumed a flow field as shown in Figure 5.29 for the derivation of his model for non-equilibrium sediment transport. The equation can be solved when the velocity components (u, w), the fall velocity w_s and the mixing coefficients ε_x and ε_z are known.

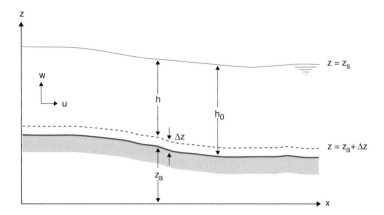

Figure 5.29. Flow field (Galappatti, 1983).

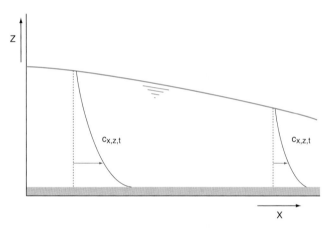

Figure 5.30. Schematization of the 2-D suspended sediment transport model.

The definitions of the symbols used in the figure are:

h = flow depth above reference level (m)

h_0 = total flow depth (m)

x, z = length coordinates (m)

u, w = velocity component in the x and z directions (m/s)

z_s = water surface elevation (m)

$\Delta z = h_0 - h$ (m)

z_a = bed elevation (m)

$\varepsilon_x, \varepsilon_z$ = sediment-mixing coefficient in the x and z directions (m^2/s)

The main concepts, on which the depth-integrated model of Galappatti is based, include:

- the horizontal diffusive transport (ε_x) and the vertical component of the velocity (w) are both ignored:

$$\left(w_s c + \varepsilon_z \frac{\partial c}{\partial z} \right)_{surface} = 0 \tag{5.121}$$

Hence, the 2D convection-diffusion equation reduces to:

$$\frac{\partial c}{\partial t} + u\frac{\partial c}{\partial x} = w_s\frac{\partial c}{\partial z} + \frac{\partial}{\partial z}\left(\varepsilon_z\frac{\partial c}{\partial z}\right) \tag{5.122}$$

- the concentration at the upstream boundary at each time step is known;
- the concentration $c_{x,z,t}$ (Figure 5.30) is presented in a depth-averaged concentration $c_{x,t}$ (Figure 5.31);
- the bed-load concentration $c_{(bed)}$ is a function of the flow and sediment parameters.

Wang and Ribberink (1986) studied the validity of the Galappatti model and they concluded that the use of the model is not suitable for large deviations of the concentration profile from the equilibrium profile. They recommended some specific requirements to be applied in the Galappatti model for the computation of suspended sediment transport. These requirements are:

- the Galappatti model is only valid for fine sediment. The factor $w_s/u*$ should be much smaller than unity; recommended values of $w_s/u*$ are between 0.3 and 0.4;
- the time scale of the flow variations should be much larger than $h/u*$;
- the length scale of the flow variations should be larger than $Vh/u*$.

where:

$u* =$ local shear velocity (m/s)
$w_s =$ fall velocity (m/s)
$\;h =$ water depth (m)
$\;V =$ mean velocity (m/s)

The boundary condition for the bed is not applied at the bed ($z = z_a$), but at a small distance Δz from the bed $z = z_a + \Delta z$. Galappatti uses in his analysis one type of bed boundary condition, i.e. the value of the concentration near the bed $c_{(bed)}$ at $z = z_a + \Delta z$. He has assumed that $c_{(bed)}$ at $z = z_a + \Delta z$ is known in terms of local flow and sediment parameters. In other words, $c_{(bed)}$ is known in advance. Equation (5.122), if integrated vertically with the pre-set boundary conditions, gives the depth-averaged equation.

$$\frac{\partial(h\bar{c})}{\partial t} + \frac{\partial(h \cdot \overline{uc})}{\partial x} = E \tag{5.123}$$

$$h \cdot \overline{uc} = \int_{Z_A+\Delta Z}^{Z_A+\Delta Z+h} uc\, dz \tag{5.124}$$

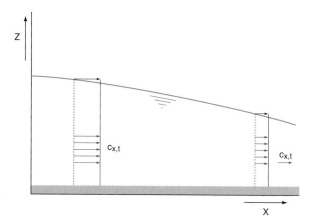

Figure 5.31. Schematization of
a depth-integrated model.

where:

$\bar{c}$ = mean concentration (m³ sediment/m³ water)
E = entrainment rate
h = depth of flow (m)
$\bar{u}$ = mean flow velocity (m/s)

For the asymptotic solution of the convection-diffusion equation the
following two assumptions are made:

$$\frac{UH}{Lw_s} = \delta \ll 1$$

$$\frac{H}{w_s T} = \delta \ll 1$$

This implies that the time taken for a particle to settle ($t = h/w_s$) is much
smaller than the time it takes to be convected along a distance L ($T = L/u$).
The general solution for the sediment concentration in non-equilibrium
conditions in terms of depth-averaged variables will be:

$$\gamma_{11}\bar{c}_e = \gamma_{11}\bar{c} + \gamma_{21}\frac{h}{w_s}\frac{\partial \bar{c}}{\partial t} + \gamma_{22}\frac{uh}{w_s}\frac{\partial \bar{c}}{\partial x} \tag{5.125}$$

$$\bar{c}_e = \bar{c} + T_A\frac{\partial \bar{c}}{\partial t} + L_A\frac{\partial \bar{c}}{\partial x} \tag{5.126}$$

with

$$T_A = \frac{w_s}{u^*}\frac{h}{w_s}\exp(f) \tag{5.127}$$

$$\frac{T_A u^*}{h} = \exp(f) \tag{5.128}$$

$$L_A = \frac{w_s}{u^*} \frac{Vh}{w_s} \exp(f) \tag{5.129}$$

$$\frac{L_A}{h} = \frac{w_s}{u^*} \frac{V}{w_s} \exp(f) \tag{5.130}$$

$$f = \sum_{i=1}^{4} \left(a_i + b_i \frac{u^*}{V} \right) \left(\frac{w_s}{u^*} \right)^{i-1} \tag{5.131}$$

where:
c_e = concentration of suspended load in equilibrium condition
c = concentration of suspended load at distance x
c_0 = concentration of suspended load at distance $x = 0$
T_A = adaptation time (s)
L_A = adaptation length (m)
x = length coordinate (m)
w_s = fall velocity (m/s)
u^* = shear velocity (m/s)
V = mean velocity (m/s)
h = water depth (m)
a_i, b_i = constants

The adaptation length (L_A) and adaptation time (T_A) are constant for uniform flow and are defined as the interval (both in length and in time) required for the mean actual concentration to approach the mean equilibrium concentration. The adaptation length and the adaptation time represent respectively the length scale and the time scale (Ribberink, 1986).

To compute the adaptation time (T_A) and adaptation length (L_A), the constants a_i and b_i are required. Galappatti (1983) used suspension parameters for natural channels and computed these constants for different ratios of $\Delta z/h_0$. The constants for T_A and L_A for $\Delta z/h_0 = 0.01$, 0.02 and 0.05 are given in Table 5.9. In steady and uniform flow, the adaptation length (L_A) and adaptation time (T_A) are constant and have straight characteristics in the x-t plane. In gradually varied flow the water depth and velocity change in each length step. Therefore L_A and T_A are not constant along the canal. Furthermore, the magnitude of $|c - c_e|$ decreases exponentially with L_A and T_A. They are defined as the interval (both in length and time) required for the actual concentration to approach the mean concentration. The adaptation length and adaptation time represent the length scale and time scale respectively.

For values of z_a/h less than 0.01, the same values of a_i and b_i will be used, because for smaller values the influence of z_a/h on the adaptation time and length is insignificant (Kerssens et al., 1979).

Table 5.9. Value of a and b for different $\Delta z/h_0$ (Galappatti, 1983).

		a_1	b_1	a_2	b_2	a_3	b_3	a_4	b_4
$\Delta z/h_0 = 0.01$	T_A	1.978	0.000	−6.321	0.000	3.256	0.000	0.193	0.000
	L_A	1.978	0.543	−6.325	−3.331	3.272	0.400	0.181	1.790
$\Delta z/h_0 = 0.02$	T_A	1.788	0.000	−5.779	0.000	2.860	0.000	0.226	0.000
	L_A	1.789	0.570	−5.783	−3.000	2.872	0.560	0.217	1.430
$\Delta z/h_0 = 0.05$	T_A	1.486	0.000	−4.999	0.000	2.306	0.000	0.247	0.000
	L_A	1.486	0.576	−5.002	−2.416	2.314	0.720	0.242	0.910

For a steady sediment flow ($\partial c/\partial t = 0$):

$$c_e - c = L_A \frac{\partial c}{\partial x} \tag{5.132}$$

After integration the equation results in:

$$c = c_e - (c_e - c_0)\exp\left(-\frac{x}{L_A}\right) \tag{5.133}$$

In the introduction it was mentioned that the rates of suspended load and bed load transport of particles larger than 0.3 mm are comparable and the more reliable sediment transport predictors in irrigation canals compute the total load (suspended and bed load). For this reason it will be essential to consider the sediment transport in non-equilibrium conditions as a whole. Although the bed load reacts instantaneously from a non-equilibrium condition to an equilibrium condition, it is assumed that the characteristic adaptation length for the bed load is the same adaptation length as for suspended load. Therefore, the total sediment transport under non-equilibrium conditions can be described by using the total sediment concentration (bed and suspended load) instead of a suspended sediment concentration.

This leads to:

$$c = c_e - (c_e - c_0)\exp\left(-\frac{x}{L_A}\right) \tag{5.134}$$

A detailed study of Galappatti's model was made by Ghimire (2003), Ribberink (1986) and Wang et al. (1986). Their research showed that there are certain limitations to this model, which most notably are:
- the error in the solution increases once the mean concentration moves relatively far away from the mean equilibrium concentration:

$$\left|\frac{c_e - c}{c}\right| \ll 1 \tag{5.135}$$

The solution is valid when the deviation of C from C_e is in the range of 0 to 50%;

- the size of the sediment is uniform which can be represented by a single fall velocity;
- w_s/u^* should be much smaller than unity (approximately 0.3 to 0.4). Figure 5.32 shows the corresponding range of sediment size (d_{50}) for flow conditions normally found in irrigation canals. Considering the maximum permissible tractive force on the canal bed under normal condition ($<5\,\text{N/m}^2$), the maximum sediment size for which this model is applicable is 0.18 mm (Ghimire, 2003);
- the maximum fall velocity w_s should be always smaller than 0.028 m/s.

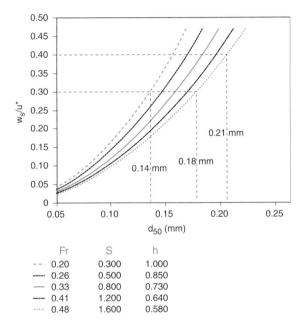

Figure 5.32. Validity range of Galappatti's model under different flow conditions (Paudel, 2010).

Fr	S	h
0.20	0.300	1.000
0.26	0.500	0.850
0.33	0.800	0.730
0.41	1.200	0.640
0.48	1.600	0.580

5.4 MORPHOLOGICAL CHANGES OF THE BOTTOM LEVEL

5.4.1 *The modified Lax method*

The one-dimensional computation assumes fixed sidewalls and the occurrence of deposited or entrained sediments on the bottom of the irrigation canals. The interrelation between the water movement and the morphological changes on the bottom can be given by the following equations:

$$\frac{dh}{dx} = \frac{S_o - S_f}{1 - Fr^2} \qquad (5.136)$$

$$\frac{\partial Q_s}{\partial x} + B(1 - P)\frac{\partial z}{\partial t} = 0 \tag{5.137}$$

These equations will be solved in sequence. First, the gradually varied flow equation is solved to determine the flow profile for given boundary conditions for water level and discharge. Details of this procedure were discussed in Chapter 3. Next, the output values of the hydraulic equation are used to solve the equation of the mass balance for the total sediment transport. The first term (dQ_s/dx) represents either the total entrainment or deposition rate between two sections (points) along the x axis of the canal. It will depend on a balance between the transport capacity and the existing sediment load along the x axis. The second term represents the net flux of sediment across a horizontal plane near the bed that will lead to a change of the bottom level (see Figure 5.32).

Several finite difference methods based on explicit and implicit schemes have been used to solve the equation of the mass balance for the total sediment transport. Cunge et al. (1980), de Vries (1987), Vreugdenhil 1989) and Vreugdenhil and Wijbenga (1982) describe the Lax, modified Lax, Lax-Wendroff and the 4-points implicit schemes as methods to solve the morphological equation. The modified Lax scheme can be used quite successfully, though it is not claimed to be the best method (Abbot & Cunge, 1982).

The modified Lax method can be expressed as:

$$z_{i,j+1} = z_{i,j} - \frac{1}{B(1-p)}\left[\frac{Q_{s_{i+1,j}} - Q_{s_{i-1,j}}}{2\Delta x} - \frac{1}{2\Delta t}[(\alpha_{i+1,j} + \alpha_{i,j})(z_{i+1,j} - z_{i,j})\right.$$
$$\left. -(\alpha_{i,j} + \alpha_{i-1,j})(z_{i,j} - z_{i-1,j})]\right] \tag{5.138}$$

The numerical scheme cannot be applied to the downstream and upstream boundaries. An adapted scheme for the downstream boundary is described by:

$$z_{i,j+1} = z_{i,j} - \frac{1}{B(1-p)}\left[\frac{Q_{s_{i+1,j}} - Q_{s_{i-1,j}}}{2\Delta x} + \frac{1}{2\Delta t}[(\alpha_{i,j} + \alpha_{i-1,j})(z_{i,j} - z_{i-1,j})]\right] \tag{5.139}$$

And for the upstream boundary by:

$$z_{i,j+1} = z_{i,j} - \frac{1}{B(1-p)}\left[\frac{Q_{si+1,j} - Q_{si-1,j}}{2\Delta x} - \frac{1}{2\Delta t}[(\alpha_{i+1,j} + \alpha_{i,j})(z_{i+1,j} - z_{i,j})]\right] \tag{5.140}$$

In which the subscripts i and j represent:

$$i = i * \Delta x$$
$$j = j * \Delta t$$

where:
Q_s = sediment discharge (m³/s)
  p = porosity
  B = bottom width (m)
Δx = distance step (m)
  z = bottom level above datum (m)
Δt = time step (m)
  α = parameter used for stability and accuracy of the numerical scheme.

The stability of the scheme is given by (Vreugdenhil, 1989):

$$\sigma^2 \leq \alpha \leq 1 \tag{5.141}$$

Accuracy of this scheme is increased if (Vreugdenhil, 1989):

$$\alpha \approx \sigma^2 + 0.01 \tag{5.142}$$

In which σ is the Courant number:

$$\sigma = NV \frac{Q_s/Q}{1 - \mathrm{Fr}^2} \frac{\Delta t}{\Delta x} \tag{5.143}$$

where:
  N = exponent of the velocity in the sediment transport equations
  Q = discharge (m³/s)
  Q_s = sediment discharge (m³/s)
  V = mean velocity (m/s)
Δt = time interval (s)
Δx = distance (m)
 Fr = Froude number

Figure 5.33 shows a schematization of the deposition or entrainment computation at the bottom of the canal and Figure 5.34 shows the calculation of the changes in the bottom level according to the modified Lax method.

5.4.2 Sediment movement

Under the conditions that the incoming sediment load into an irrigation canal with uniform flow is equal to the equilibrium transport capacity of the canal and there is no entrainment of any sediment from the bottom,

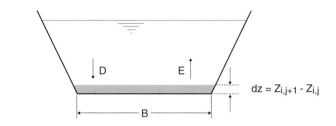

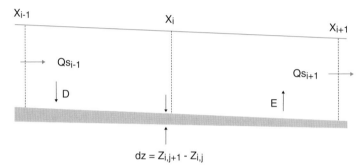

Figure 5.33. Schematization of computations of changes on the bottom level.

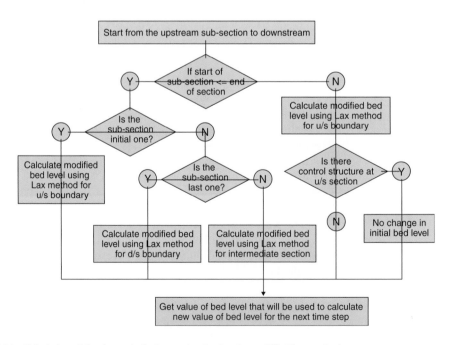

Figure 5.34. Calculation of the change in the bottom level using the modified Lax method.

then the sediment will be transported in equilibrium and there will be no deposition or erosion. If the inflowing water carries more sediment than the equilibrium transport capacity then the extra sediment will be deposited in the canal until a new equilibrium in the flow conditions is attained.

Under uniform flow conditions, when the inflowing water carries less sediment (Q_s) than the equilibrium sediment transport capacity Q_{se}, two situations might occur depending on whether there is any sediment motion on the bottom or not. Motion of sediment is evaluated in terms of the mobility parameter θ and the critical mobility parameter θ_{cr}.

- In the first situation ($\theta > \theta_{cr}$), entrainment of particles and an increasing sediment transport will occur along the canal until adaptation to the equilibrium sediment transport capacity has been attained. The readjustment of the actual sediment concentration to the equilibrium condition will take some distance and time. This distance is called the adaptation length and the time the adaptation time. Initially sediment is picked up directly downstream of the point where the flow enters the erodible region. This process will cease after some time when the bottom slope near the head becomes flatter and less sediment is picked up. Then the pickup of sediment will move to a point that is further downstream and the process is continued until the whole canal reach has attained a sediment transport that is in equilibrium with the inflowing sediment load.
- In the second case ($\theta < \theta_{cr}$), without any motion of sediment along the bottom ($\theta < \theta_{cr}$), the actual sediment load (Q_s) is conveyed without changes; the sediment transport will continue under non-equilibrium conditions.

However, the operation of irrigation structures to control and distribute irrigation water will result in most cases in non-uniform flow and these flow conditions will consequently change the sediment transport capacity with distance and in time. A sediment transport that is not in equilibrium may occur under both uniform (Figure 5.35) and non-uniform flow conditions. For gradually varied flow conditions a distinction between backwater and drawdown effects can be made (see Figure 5.36). In these flow conditions continuous erosion or deposition takes place even after an initial adjustment of the sediment concentration to the equilibrium concentration. The sediment transport capacity of the water keeps on changing with distance and time. The flow parameters adapt almost instantly to a change in conditions and the change in flow pattern is smooth. However, the adaptation of the sediment concentration requires some time to adjust to the new conditions. For a backwater there may be a continuous deposition or first erosion and then continuous deposition or a non-equilibrium flow depending upon the actual sediment concentration, sediment transport capacity and the type of bed material. For a drawdown condition there may be first deposition and then erosion, or continuous erosion or non-equilibrium conditions.

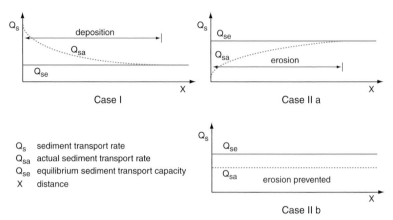

Figure 5.35. Sediment transport
for uniform flow.

Figure 5.36 shows the effect of a backwater profile on the sediment transport in a canal.

- For Q_s larger than Q_{se} deposition will occur to adapt to the equilibrium sediment transport capacity ($Q_{se} = Q_s$). From that point onwards a continuous deposition in the downstream direction may be expected.
- For Q_s smaller than Q_{se}, the actual sediment transport can either remain constant along the canal ($\theta < \theta_{cr}$) or increase up to Q_{se} ($\theta > \theta_{cr}$). The flow will attempt to attain equilibrium.

The effect for a drawdown profile is shown in Figure 5.36.

- In a drawdown flow and for Q_s larger than Q_{se} deposition will occur up to the equilibrium condition ($Q_{se} = Q_s$); the water will pick up sediment from the bed. From that point onwards a continuous entrainment is expected.
- When Q_s is smaller than Q_{se} the former either remains constant ($\theta < \theta_{cr}$) or increases to reach Q_{se} ($\theta > \theta_{cr}$). From that point onwards a continuous entrainment in the downstream direction will occur.

Figure 5.36 shows some of the steps in the sediment behaviour for the two gradually varied flow situations in an irrigation canal. They also illustrate the procedure to compute the sediment mass balance after one time step.

The sediment movement in irrigation canals is not simple once flow control structures are introduced. The standard design practices that include the sediment transport aspect are not capable of predicting the sediment transport behaviour, including erosion and sedimentation. The design assumes a steady and uniform flow condition and the design discharge is generally the maximum discharge, which is not released during most of the time. Once the flow is less than the design discharge the sediment transport capacity will decrease even for the new uniform flow

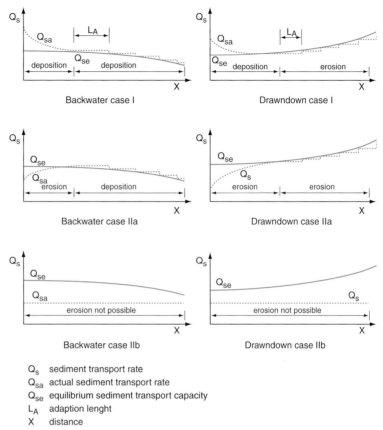

Figure 5.36. Sediment transport for gradually varied flow.

Q_s sediment transport rate
Q_{sa} actual sediment transport rate
Q_{se} equilibrium sediment transport capacity
L_A adaption lenght
X distance

condition. To maintain the water level at the set points for flows smaller than the design values, the gates are lowered, which will create a backwater effect resulting in non-equilibrium sediment transport conditions.

Therefore the design approach should be modified in such a way that the actual sediment movement scenario is reflected. Only then will the designed canal result in an improved sediment transport under changing flow conditions.

5.5 CONCLUSIONS

Applications of the existing sediment transport concepts under flow conditions and sediment characteristics prevailing in irrigation canals were used to evaluate the suitability of those concepts. In this way any increase in the unavoidable uncertainties and inaccuracies during the computations of the sediment transport can be kept to a minimum. From the analysis of the sediment transport concepts for the conditions prevailing in irrigation canals the following conclusions can be drawn:

General

- there are several theories to estimate friction factors and (equilibrium and non-equilibrium) sediment transport rates. These theories rely on field and laboratory data, but they are not able to predict very accurately the friction factors and sediment transport rates in irrigation canals;

Bed forms

- for the lower flow regime in irrigation canals all types of bed forms (ripples, mega ripples and dunes) can be expected;
- the van Rijn method describes the bed forms more accurately than other methods; more than 75% of the observed bed form types were *well predicted* by the van Rijn method;

Friction factor predictors

- the van Rijn method to predict the friction factor due to the grains and bed forms gives good results when compared with measured values of the friction factor. This method was able to predict more than 90% of the measured values within an error band of 30%;
- several existing friction factor predictors consider only the bottom friction. The sidewalls of non-wide canals have an important impact on the overall friction and therefore a weighed composite friction factor is required. The proposed method for predicting the equivalent roughness in a trapezoidal canal with different roughness along the bottom and sidewalls performs better than the existing methods. The method predicted more than 90% of the measured values as *well predicted* and resulted in the minimum value of the standard error and the narrowest range of variation of the predicted values;
- the existing methods for predicting the equivalent roughness in a rectangular canal with different roughness along the bottom and vertical sidewalls cannot be explicitly applied. Mendez proposed a new method to estimate the effective roughness in a rectangular canal with composite roughness. The method predicted more than 95% of the measured values within an error band of 15%;
- the roughness of the side slopes is due to the type of material and the vegetation, which may be present at the beginning or likely to grow during the irrigation season, depending upon the type of maintenance. The roughness has been divided into two categories, namely sides without any vegetation and sides that are covered with vegetation. The second category has been further divided into various weed factors from densely grown to 1.0 ideally clean. For ideally maintained conditions a weed factor may vary from 0.9 to 1.0. For fully-grown vegetation the weed factor ranges between 0.1–0.3;
- maintenance conditions are divided into three possible maintenance scenarios: no maintenance, well maintained and ideally maintained. For poorly maintained canals, at each time step starting from the beginning

the effect of vegetation is accounted for in the roughness calculation. The maximum value occurs if the simulation period equals or exceeds the total growing period of the vegetation. For well-maintained canals, the effect of vegetation on the roughness is considered from the beginning. Weed factor 1 is taken for the initial stage and it increases linearly up to the periodic maintenance time. Méndez (1998) defined the periodic maintenance interval at approximately two months; however it depends upon the type of vegetation in terms of total growing time and its effect on roughness;

Sediment transport predictors

- based on the typical flow conditions and the sediment sizes usually encountered in irrigation canals, the sediment will be transported as bed load and suspended load, separately or combined. Therefore the sediment transport predictors for irrigation canals should be able to compute both types of sediment transport;
- existing predictors of the sediment transport capacity of wide canals do not take into account the geometry and its effect on the velocity distribution over the cross section; they do not describe the sediment transport in irrigation canals in a reliable or realistic way. The best prediction methods give only 60% of the measured values as *well predicted* within an error band of 100%;
- the Ackers-White and the Brownlie predictors provide more accurate estimates of the sediment transport capacity under the prevailing flow conditions and sediment characteristics in irrigation canals.

CHAPTER 6

SETRIC, a Mathematical Model for Sediment Transport in Irrigation Canals

6.1 UPSTREAM CONTROLLED IRRIGATION SYSTEMS

6.1.1 *Introduction*

Clogging of the structures that supply water to secondary and tertiary units, and sedimentation of the canal network are some of the main problems in the operation and maintenance of irrigation systems. A high annual investment is required for rehabilitation and maintenance to keep the irrigation systems suitable for their purpose. To reduce this expense and to preserve and sometimes improve the performance of an irrigation network, the sediment transport should be properly estimated in terms of time and space. An accurate prediction of the sediment deposition along the entire canal network during the irrigation season will contribute to an improved operation of the canals in such a way that the irrigation needs are fully met and at the same time a minimum deposition might be expected.

Although it is difficult to predict the quantity of sediment that will be deposited in irrigation canals (Brabben, 1990), numerical modelling of sediment transport offers the possibility to predict and evaluate the sediment transport under general flow conditions (Lyn, 1987). A mathematical model, which includes the latest sediment transport concepts for the specific conditions of irrigation canals, will be an important and helpful tool for the designers and managers of these systems. This chapter will describe the computer model SETRIC, which has been in development at UNESCO-IHE since 1994 and which includes some of the specific characteristics of sediment transport in irrigation canals, as discussed in the previous chapter.

The sediment transport model SETRIC (Méndez, 1998; Paudel, 2002) is a one-dimensional model where the water flow is schematized as a quasi-steady and solved as a gradually varied flow. (The model can be found on the Internet.) The sub-critical flow profile is determined by the predictor-corrector approach. Roughness on the canal bed and side slopes is calculated separately. The model computes the roughness on the bed using the van Rijn (1984c) method, which is based on flow conditions and bed form and grain related parameters such as bed form length, height and sediment size. Then the equivalent roughness is computed taking into account the sidewall effect as proposed by Méndez (1998) and Paudel (2010).

The sediment continuity equation is solved numerically by using the Modified Lax scheme. A depth-integrated convection-diffusion model (Galappatti, 1983) is applied to predict the actual sediment concentration at any point under non-equilibrium conditions. For the prediction of the total sediment transport under equilibrium condition an option is available to select one of the three sediment predictors, namely Ackers-White (HR Wallingford, 1990), Brownlie (1981b) and Engelund-Hansen (1967). The predicted sediment transport capacity is corrected for the B/y ratio and side slope for non-wide irrigation canals.

6.1.2 *Water flow equations*

The schematization of the water flow in irrigation canals will have to consider two types of aspect, namely the operational aspects and the sediment transport aspect. The operational aspects include the situation that the uniform water flow will become non-uniform and unsteady due to the changing nature of the irrigation demand and the operation of the gates to meet this demand and to maintain the water supply level for a proper irrigation of the fields. The sediment transport aspect covers the changes in the sediment morphology, which are much slower than the changes in the water flow with time and space.

The water flow in irrigation canals during an irrigation season and moreover during the lifetime of the canals is not constant. Seasonal changes in crop water requirement, water supply and variation in size and type of the planned cropping pattern are frequent events during the lifetime of an irrigation canal. The operation of gates to adjust to the variation in supply is normally gradual to avoid the generation of waves. It is common practice to allow sufficient time to move from one steady state to another steady state by operating the gate in multiple stages. Moreover, the Froude number in irrigation canals is normally small (<0.4) to maintain a stable water surface and small disturbances in bends and transitions (Ranga Raju, 1981). For small Froude numbers (Fr < 0.6–0.8), the celerity of the water movement (c_w) is much larger than the celerity of the bed level change (c_s) (de Vries, 1975): therefore the water flow and the sediment transport calculations can be uncoupled. For the non-uniform flow situations the water flow can be treated as a gradually varied flow.

Water flow in irrigation canals will be schematized as quasi-steady in which the governing equations are represented as:

Continuity equation:

$$\frac{\partial Q}{\partial x} = 0 \qquad \therefore Q = \text{constant} \tag{6.1}$$

Continuity equation at confluences and/or bifurcations:

$$\sum \text{inflow} = \sum \text{outflow}$$
$$Q \pm q_1 = 0 \tag{6.2}$$

Dynamic equation:

$$\frac{dh}{dx} = \frac{S_o - S_f}{1 - \text{Fr}^2} \tag{6.3}$$

$$\text{Fr}^2 = \frac{Q^2 B_s}{gA^3} \tag{6.4}$$

Several methods are available to solve the dynamic equation of gradually varied flow for prismatic canals. Henderson (1966), Chow (1983), Depeweg (1993) and Rhodes (1995) present a comprehensive description of the available methods, including graphical integration, direct integration, direct step, standard step, the Newton-Raphson solution and the predictor-corrector method. A summary of these methods is given in Chapter 2.

6.1.3 *Sediment transport equations*

The analysis of the sediment transport process is based on the following assumptions:
- the sediment particles can be characterized by a single representative size;
- the size and gradation of the sediment remain constant along the whole length of the canal and throughout the irrigation season;
- the canal bed is composed of the same material as that of the inflowing sediment.

The numerical solution of the one-dimensional sediment equations that include the friction factor predictor, continuity equation for sediment and the sediment transport predictor (see Chapter 5) requires boundary conditions for the hydraulic and sediment transport calculations.
These conditions are:
- the geometrical variables of the canal: bottom width, side slope and level of the bottom;
- the flow conditions along the entire canal during a time step: discharge, velocity, roughness, water depth and slope of the energy line;
- specific confluences and bifurcations can be incorporated by applying continuity for water flow and sediment load;
- the characteristics of the incoming sediment, namely the sediment load (ppm) and sediment size d_{50} at the upstream boundary; there will be no entrainment of the original bottom material;

– the sediment transport rate along the entire canal;
– the changes in bottom level and/or bottom width.

In order to compute the sediment transport along the entire canal, it is necessary to consider how the incoming sediment load c_0 adapts to the equilibrium sediment transport of the canal reaches; this is given in terms of the sediment concentration c, which follows from Gallapatti's depth-integrated model. Galappatti (1983) has developed a depth-integrated suspended sediment transport model based on an asymptotic solution for the two-dimensional convection equation in the vertical plane (see Chapter 5). SETRIC uses the diffusion model developed by Galappatti.
    This depth-integrated model solves the convection-diffusion equation with the following assumptions:
– diffusion terms other than vertical are neglected;
– concentration is expressed as the depth-averaged concentration.

As mentioned the Galappatti model has two advantages over others; firstly, no empirical relation has been used during the derivation of the model and secondly, all possible bed boundary conditions can be used (Wang and Ribberink, 1986). The value of c follows from:

$$c = c_e - (c_e - c_0)e^{-x/L_A} \tag{6.5}$$

$$L_A = f\left(\frac{u_*}{v}, \frac{w_s}{u_*}, y\right) \tag{6.6}$$

where:
  c = actual concentration (ppm)
  c_e = equilibrium concentration (ppm)
  c_0 = initial concentration at $x = 0$ (ppm)
  x = distance along the canal (m)
  L_A = adaptation length (m)

For the determination of the actual sediment concentration in the x-direction of the canal, the values of the variables c_0, L_A, Δx and c_e have to be known. The first, the initial concentration c_0, does not depend on the local flow condition, but instead depends on the inflowing water and sediment and on whether a sediment trap is located at the head of the irrigation network. At the boundaries between canal reaches, the sediment load passing through the downstream boundary of the upstream reach will become c_0 for the next canal reach.
    In a gradually varied flow the values of c_e, y, v, u_* and f are functions of distance x and time t. These variables may be known in advance at any point along the canal network if the flow equations are solved first (using the uncoupled technique). That means that c_e, y, v, u_* and the

dimensionless parameters u_*/v and w_s/u_* are given for any point of the canal ($i = 0, 1, 2$ to n).

These variables are determined according to the following procedure:

- the Δx value is fixed according to the required degree of accuracy for the numerical solutions and the need for representing the adaptation of the actual non-equilibrium condition to the sediment transport capacity (the equilibrium condition) of the canal. The Δx value should be much smaller than the adaptation length (L_A) of the actual sediment transport to the sediment transport capacity (equilibrium concentration);
- computation of the L_A-value. For the local flow conditions the value of w_s/u_*, u_*/v and the water depth are known in advance. The maximum values of the parameter w_s/u_* should satisfy the requirements for the validity of the depth-integrated model (Ribberink, 1986);
- once the motion of sediment has been initiated, the values of the de Chézy coefficient can be estimated depending on the type of roughness along the wetted perimeter of the canal by:

$$C = 18 \log \frac{12R}{k_s} \quad \text{for single roughness} \tag{6.7}$$

$$C'_e = 18 \log \frac{12R}{k_{se}} \quad \text{for composite roughness} \tag{6.8}$$

- the value of the depth-averaged equilibrium concentration C_e will be determined by one of the sediment transport predictors, for example Ackers-White, Brownlie or Engelund-Hansen. The predictors compute the sediment transport per unit width (q_s), which is determined by the local flow conditions and the sediment properties. The total sediment transport across the whole canal section Q_{se} is calculated by:

$$Q_{se} = \alpha B q_s \tag{6.9}$$

- next, the equilibrium concentration c_e is calculated by:

$$c_e = \left(\frac{\rho_s}{\rho} \frac{Q_s}{Q} \right) * 10^6 \text{ (ppm)} \tag{6.10}$$

- some internal conditions along the canal can be taken into account for the computation of the sediment distribution either at boundaries between canal reaches or at branches (bifurcations or confluences). Application of the continuity for the water flow and for the sediment transport is effective when changes either in bottom width or bottom level, or at bifurcations or confluences have to be simulated by the model. The distribution of water and sediment at bifurcations will depend on the local flow pattern at the branches.

6.1.4 *General description of the mathematical model*

The mathematical model SETRIC simulates the water flow, sediment transport and changes of bottom level in an open irrigation network with a main canal and several secondary canals with or without tertiary outlets. Various flow conditions along the canal network and during the irrigation season can be simulated. Figure 6.1 shows the flow diagram for calculating the change of the bottom level in a canal reach during one time step. Flow diagrams of the computer program SETRIC for the water flow and sediment transport calculations in a canal reach are shown in Figures 6.2 and 6.3. The specific module for downstream controlled irrigation systems will be discussed in Section 6.2.

The background for the hydraulic and sediment transport computations was described in the previous chapters. The general framework of the computer program consists of:

Hydraulic aspects
– The water flow can be modelled as a sub-critical, quasi-steady, uniform or gradually varied flow (backwater as well as drawdown curves). The water profiles for the subcritical, gradually varied flow include: H2, M1, M2, C1, S1 and A2;
– The water flow is in open canals with a rectangular or trapezoidal cross section; only friction losses are considered: no local losses due to changes in the bottom level, cross section or discharge will be taken into account. Seepage losses are also not included in the model;
– The canal sections are characterized by the following geometrical dimensions:
  • length (l) presents the length of a canal reach (m);
  • bottom width (B);
  • side slope m (1 vertical : m horizontal);
  • the roughness is defined by the equivalent roughness coefficient (k_s); the model computes the total friction factor from the composite roughness of the cross section (See Appendix F);
  • the distance is measured along the canal axis; coordinates give the location of a canal reach; the most upstream boundary is defined as $x = 0$ m;
  • bottom slope (S_o in m/m);
  • bottom elevation above a reference level (datum) at the beginning of each canal section (z_b).

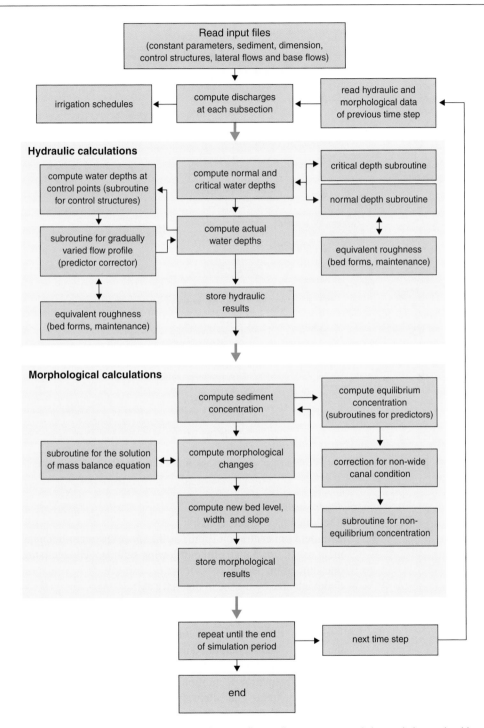

Figure 6.1. Flow diagram of SETRIC for calculating the water flow, sediment transport and changes in bottom level in main and/or lateral canals (after Paudel, 2002).

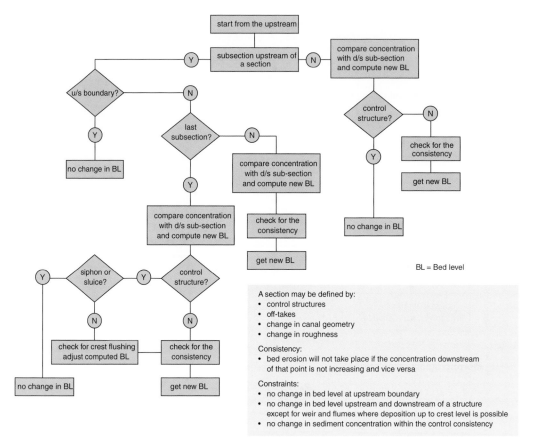

Figure 6.2. Flow diagram for SETRIC for calculating the water flow in main and lateral canals during a time step (after Paudel, 2002).

Sediment aspects

The sediment is characterized by:

– Sediment concentration (ppm) at the most upstream boundary of the main canal; no change of the inflowing sediment concentration during one simulation period;

– Sediment size by the mean diameter d_{50};

– No sediment deposition in the structures, the sediment inflow is equal to the sediment outflow;

– Variations of the roughness conditions over time are incorporated in the model; sedimentation during the irrigation season will induce the development of bed forms, which depend on the flow conditions (different flow conditions will produce different types of bed form). The total equivalent friction factor is computed for every time step and for each flow condition in each cross section of the schematization;

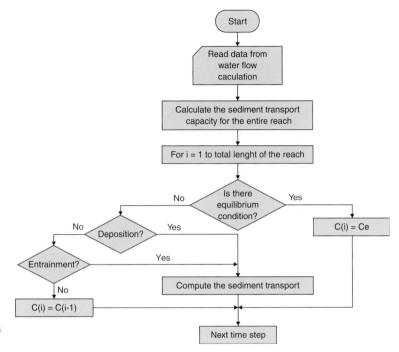

Figure 6.3. Flow diagram of SETRIC for calculating the sediment transport in the main canal and lateral canals during a time step.

– The model simulates the total sediment load and not the bed load and suspended load separately. If a structure has a raised crest then it might trap a part of the bed load upstream of the structure. Only the whole suspended load part will move downstream of the structure. Out of the total load the suspended load-carrying capacity of the canal is calculated using the following relation (Méndez, 1998):

$$q_{s,s} = \frac{q_{s,t}}{\left[\dfrac{5}{12} * \left(\dfrac{d_{50}}{y} \right)^{0.2} * D_*^{0.6} + 1 \right]} \qquad (6.11)$$

where:
$q_{s,s}$ = suspended part of the sediment transport rate (m²/s)
$q_{s,t}$ = total load transport rate predicted (m²/s)
d_{50} = mean sediment size (m)
y = water depth (m)
D_* = dimensionless grain parameter

– In the case of lateral inflow into the system the downstream concentration c_2 will change accordingly, depending upon the incoming concentration and discharge. Continuity for the water flow and sediment transport gives: $Q_2 = Q_1 + Q_i$ and $c_2 Q_2 = c_1 Q_1 + c_i Q_i$.

The change is given by:

$$c_2 = c_1 \frac{Q_1}{Q_2} + c_i \frac{Q_i}{Q_2} \tag{6.12}$$

where:
c_1 = sediment concentration in the canal upstream of the inflow (ppm)
c_2 = sediment concentration after the inlet (ppm)
c_i = concentration of inflow (ppm)
Q_1 = discharge before the inlet (m³/s)
Q_i = inflow discharge (m³/s)
Q_2 = discharge after the inlet (m³/s)

For an outflow the continuity for the water flow gives $Q_1 = Q_2 + Q_i$ and the sediment concentration after the offtake point is given by:

$$c_2 = \frac{c_1 Q_1}{(Q_2 + f_d * Q_o)} \tag{6.13}$$

where:
Q_1 = discharge before the offtake (m³/s)
Q_o = offtake discharge (m³/s)
f_d = sediment distribution ratio at the offtake = c_o/c_2

Irrigation aspects
– The model of the irrigation network can be most simply composed of a main canal and secondary canals with tertiary outlets. Each canal is divided into several reaches or sections;
– The model can include changes in the bottom level at the upstream boundary of a canal section;
– Control sections can be set at the downstream end of the main canal or secondary canals. The type of structure located at the downstream end of a section sets the water level, both for upstream and downstream control: the first one sets the upstream water level and the other the downstream water level;
– The network might include lateral inflow or outflow; these lateral flows should be located at the end of a canal section for upstream con-trolled systems and at the downstream side of the gate for downstream controlled irrigation systems;
– The flow control structures that can be incorporated include:
  • *overflow type*: crest width, crest level;
  • *undershot type*: width and height of the rectangular opening;
  • *submerged culverts and inverted siphons*: number and diameter of pipes;
  • *flumes*: the upstream head-discharge relationship;

- *drops*: incorporated as a difference in bottom level at the boundary between two reaches;
- *AVIO* and *AVIS gates* for downstream control.

- The variation in crop water requirement during the irrigation season can be incorporated and mainly depends on the climate, cropping pattern, stage of the crops, leaching requirement and water losses. The growing season is divided into four stages depending on the crop development and climate (FAO, 1998). The water supply in the model can be attuned to the varying water requirement and follows the changes in area and time;

- The irrigation schedule can include rotational turns of water supply (intermittent) to the secondary canals or tertiary off takes; when there is no flow in the canals no sediment deposition will occur in the model during the turn-off time;

- The composite roughness for a cross section with obstruction is derived from the initial roughness of the whole cross section, depending on whether there is either a single or composite roughness or based on the type of roughness of the bottom and sides;

- The variation of the weed factor during a certain period will depend on the type of maintenance. Maintenance activities in view of the weed growth are referred to by an obstruction degree. The effect of maintenance on the roughness is expressed by:
  - ideally maintained: negligible obstruction degree over time;
  - well maintained: a maximum obstruction degree of 10% is assumed;
  - poorly maintained: more than 75% obstruction degree is assumed;
  - the actual roughness is calculated by using the weed factor F_w.

6.1.5 *Input and output data*

Input data

For the computation of the water flow and sediment transport, the program requires several types of input data, which include:

- *Simulation period*: the characteristics of the irrigation season include:
  - number of periods in which the irrigation season is divided;
  - type of maintenance to be expected during each period of the irrigation season;
  - number of days for each period and the number of irrigation hours per day.

- *Canal dimensions*: the geometrical dimensions of main and secondary canals contain:
  - for each canal section: the number of sections, type of roughness (de Chézy, Manning or Nikuradse);
  - for each canal section: the location from the upstream boundary, length, side and bed slope, width and elevation of the bottom; roughness coefficient.

- *Main and lateral discharges*: the total simulation period; the irrigation flows (inflow (+) and outflow (−)) during each period of the simulation; the discharge entering the main canal for each period together with the lateral flows at the furthest upstream boundary of each canal section and for each period; no seepage, operation or other losses are included in the model;
- *Sediment data*: mean sediment concentration (ppm) and mean diameter d_{50} of the sediment particles entering the main canal; the sediment predictor and the sediment inflow constant or in time series; SETRIC computes the sediment concentration flowing to the secondary canals;
- *Control sections*: the control section at the downstream end of the main canal and at each lateral has to be specified and they include the water depth and the location of each control section. Control structures at the boundaries between canal reaches are specified by the type of structure (overflow, undershot, Crump weir, flume, siphon or AVIS gates); and its main hydraulic characteristics (constant or time dependent). The model computes the upstream and downstream water level at a control section; the latter will act as a control level for the next reach;
- *Canal bottom:* the model assumes that the original canal bottom has a hard layer; no erosion will take place below this original bottom level;
- *Embankments:* height of embankments or structures is not included in the model; the hydraulic simulation will continue without overtopping in a virtual infinite cross-section;
- *Weed factors and other parameters:* the weed factors are determined by their growing period, periodic maintenance, maximum weed factor and the weed effects at three growing stages; other parameters concern the sediment division at off takes, minimum roughness height; specific water and sediment properties and modelling parameters (time and length step). (See Appendix F).

Output data

The computer program is able to present the results of the hydraulic and sediment transport calculations in tables or graphs depending on the selected option. Moreover, the results can be presented on screen, on paper or saved in files. The program includes the following tables:
- *hydraulic results*: related to the water flow: the initial and final water flow, including critical, normal and actual water depth, discharge, bed width, bottom slope and de Chézy coefficient;
- *concentrations*: table for the initial and final sediment transport results with the actual water depth and level, equilibrium concentration and actual concentration, initial bed level and modified bed level and deposition volume per length step for the entire canal.

The program contains the following figures for the main canal and separately for the laterals:
- Initial water flow (actual, normal and critical depth and bottom level);

- Final water flow (actual, normal and critical depth and modified bottom level);
- Initial sediment concentration along the canal (actual and equilibrium);
- Final sediment concentration along the canal (actual and equilibrium);
- Initial and final bed levels along the canal;
- Local and accumulated deposition in the canal at the end of the simulation.

6.2 DOWNSTREAM CONTROLLED IRRIGATION SYSTEMS

6.2.1 *Background*

Only a few sediment transport models for open channels can be applied to irrigation canals. There are some commonly used models for steady and unsteady state flow simulations in irrigation canals such as RootCanal (Utah State University, 2006), SIC (Baume et al., 2005) and CanalMan (Merkley, 1997), but only the SIC model, out of these three models, is capable of modelling the sediment transport in irrigation canals.

Flow and sediment transport simulations in demand-based systems consider automatic flow controllers as an integral part of the models. Generally most mathematical models do not have an automatic flow control module and the application of these models for automatically controlled irrigation systems becomes difficult. Additionally, these simulations need fully dynamic models which can solve the Saint-Venant equations, which takes a lot of time during simulations.

These factors together make simulation complicated and too difficult for designers and canal managers, particularly in canal automation. A simple model which would be able to simulate flow and sediment transport in supply and demand-based irrigation canals with fixed and automatic flow operations would help designers and canal managers. Therefore SETRIC has been further developed to cope with these problems of flow and sediment transport simulations in demand-based irrigation canals. The concept of this model was developed by Méndez (1998) with the focus on sediment transport modelling in irrigation canals. Later, Paudel (2002; 2010) developed and tested the model for supply-based irrigation canals. Munir (2011) improved SETRIC with the incorporation of a module for downstream controlled systems.

6.2.2 *Downstream control considerations*

The approach of the hydraulic aspects of the downstream controlled irrigation systems is somehow different from the upstream controlled systems. The hydraulic models developed for upstream control need some modifications to be applied for downstream control systems. In upstream control the water levels in a canal drop accordingly with a reduction in flow, but in downstream control systems the water levels must increase with a decrease in flow. The upstream controlled systems are usually manually operated, while the downstream controlled systems are mainly automatically controlled. In upstream controlled systems the set point (target water level) is set at the upstream side of the flow and water level control structures, while in downstream controlled systems these are fixed at the downstream side of the regulating structures. Therefore the flow control algorithms developed for upstream control cannot be directly applied to downstream control.

Besides, downstream controlled systems are generally applied to demand-based irrigation. The inflowing discharge at the canal headworks remains variable in response to the operation of the secondary or tertiary irrigation off takes. In steady state models the reduction of the flow at the head causes a drop in the downstream water levels in the irrigation canals and vice versa, which is in principle contradictory to the flow control methods in downstream controlled systems. In downstream controlled systems any decrease in canal flow causes an increase in the canal water levels and vice versa. This increase provides water storage in the canal, which is used to diminish the effect of canal filling or response time when an offtake is opened again after closure. This water storage provides immediate supply to the secondary or tertiary system without decreasing the flows to other offtaking canals, which may otherwise be faced in the case of manually upstream controlled systems.

In SETRIC this problem has been solved and there is no need to give the incoming flow as a function of time at the canal headworks; only the design discharge is needed at the canal headworks. Then according to the operation of the secondary or tertiary offtakes the flow is automatically adjusted. The water levels in the canal will automatically respond to flow variations in the system and increase or decrease accordingly. As the inflow is generated automatically, it eliminates the need for manually operated flow control at the headworks for a steady state. Incorporation of these options in SETRIC has resulted in a simple and easy tool to apply for flow and sediment transport modelling in upstream and downstream controlled irrigation canals.

The main objective of this section is to describe the functioning of automatically downstream controlled irrigation systems and to focus on the associated complications in hydraulic and sediment transport modelling by a simple steady state model.

6.2.3 *Aspects of the downstream control module*

The mathematical model SETRIC simulates the water flow, sediment transport and changes of the bottom level in downstream-controlled irrigation systems in the same way as in open upstream-controlled irrigation networks. Many flow conditions along the canal network and during the irrigation season are the same in upstream and downstream-controlled irrigation systems. Some specific adjustments in the module for downstream-controlled irrigation systems will be discussed in this section; as discussed in Section 6.1.4 the general framework of the computer program consists of:

Hydraulic aspects
– The water flow can be modelled as a sub-critical, quasi-steady, uniform or gradually varied flow (backwater as well as drawdown curves);
– The irrigation canals have a rectangular or trapezoidal cross section; only friction losses are considered: no local losses due to changes in the bottom level, cross section or discharge will be taken into account; also seepage losses are not included;
– The distance x is measured along the canal axis; at the most upstream boundary x is 0 m; the bottom slope S_o (m/m) and the bottom elevation above a reference level.
– The roughness is defined by the equivalent roughness coefficient (k_s); the total friction factor follows from the composite roughness of the cross section and the weed factor (see Section 6.1.4);
– The flow control structures in this module include AVIO and AVIS gates. Downstream controlled canals are generally equipped with AVIO and AVIS gates that control the water levels at the headworks and in the canal reaches. These hydro-mechanical self-operating gates maintain a constant water level at their downstream side. The AVIS gate have a free surface flow and AVIO gates operate under orifice conditions. Two types of AVIS/AVIO gates are available, namely the High Head and the Low Head type. The high head gates are usually employed in irrigation canals with narrow canal cross sections. The choice between the open type and the orifice-type is determined by the maximum head loss that will occur between the upstream and downstream water levels. Ankum (2004) gives several details of the AVIO and AVIS gates, including the available gate types, their index and dimensions. The manufacturer of the AVIS/AVIO gates (GEC Alsthom, previously Neyrtec, Neyrpic from France) provides head loss charts, which present the relationship between the discharge and the upstream water level above the gate axis. Both gates are water level and not discharge regulators; they are not water-tight. If the flow has to be shut off completely, the gate needs some stop-logs to close the structure.

The discharge computation for the AVIS/AVIO gates is a two-step procedure. Firstly the gate opening follows from the downstream water level and then this opening is used to compute the discharge. Munir (2011) has given the head discharge relations that are incorporated in SETRIC.

The design steps include:
– the gate opening is calculated from the downstream water level (set point) and the design and tuning parameters;
– then the maximum angle of the gate opening is determined, which gives the relative vertical gate opening;
– the gate opening is used to find the head-discharge relation of the gate.

The computation procedure for the SETRIC module for downstream control is as follows:
– the hydraulic computations start downstream of the most downstream gate for the given discharge and water level;
– the downstream water level results in the difference (the actual decrement) between the elevation of the gate axis and the downstream water elevation;
– the established actual decrement gives the gate opening angle;
– the gate opening angle gives the actual gate opening;
– this gate opening and the discharge lead to the water level upstream of the gate;
– the upstream water level is then the boundary for the next upstream canal reach;
– the water surface profile is computed from the tail structure to the first upstream cross regulator taking into account factors such as offtake withdrawal. The gradually varied glow equation is then solved by the predictor-corrector method;
– the procedure is continued until the upstream end of the first, most upstream canal reach;
– after the hydraulic calculation follow the sediment transport computations that will start from the upstream boundary and go in a downstream direction to the end of the canal;
– then in the next time step the same procedure will be repeated.

The hydraulic and sediment transport computations are performed for every time step. In one step first the hydraulic computations are performed and next the sediment transport computations are done according to the hydraulic conditions at the end of that time step. The flow changes will affect the sediment transport behaviour and the sediment transport will influence the canal hydraulics in the next time step.

Sediment aspects
The sediment aspects are fully comparable with the aspects for upstream-controlled irrigation systems (see Section 6.1.4).

Irrigation aspects
– The model of the irrigation network can be composed of a main canal and secondary canals with tertiary outlets. Each canal is divided into several reaches or sections;

– The model can include changes in the bottom level at the upstream boundary of a canal section;
– The control sections are set at the downstream end of the main canal or secondary canals; the structure at the downstream end sets directly the water level downstream of the gate for the hydraulic computations;
– The network might include lateral inflow or outflow; these flows are located at the downstream side of the gate;
– The variation in water supply can be attuned to the varying water requirement and follows the changes in area and time.

6.2.4 *Input and output data*

Input data
The input and output data for the computation of the water flow and sediment transport in downstream control are comparable to the input and output data previously mentioned for the upstream control (Section 6.1.5):
– *Simulation period*
– *Canal dimensions*
– *Canal bottom*
– *Sediment data*

Some differences are found in:
– *Design discharge*: the discharge at the upstream boundary;
– *Water level*: the tail water level downstream of the gate (downstream pond level); the tail water level will not change during one simulation period.
– *Main and lateral discharges*: the location of the inflow (+) and outflow (−) is at the downstream side of the automatic gates; for downstream control the tail outflow (−) and the tail water depth are required;
– *Control sections*: the water level control at the downstream side of the automatic gates includes the water level and the location of the structure. The automatic gates are specified by one of the three types of AVIS gates (200/375, 180/335 and 160/300) and the number of gates; other data include the outflowing discharge, downstream pond level and the decrement over the gate;
– *Downstream control*: the model computes the upstream water level from the downstream water level; this upstream level will act as the control level for the hydraulic computations of the next upstream canal-reach; the automatic gate at the upstream side of the reach will adjust the discharge to maintain the set point at the downstream side of the gate.

Output data
The computer program presents the results of the water flow and suspended sediment transport calculations in tables, graphs or in files

depending on the selected option. The program encompasses the following tables:
- Hydraulic results: results related to the water flow: the initial and final water flow, including critical, normal and actual water depth, discharge and de Chézy coefficient;
- Concentrations: table for the initial and final sediment transport results with the water depth, equilibrium concentration and actual concentration, initial bed level and modified bed level and deposition volume per length step for the entire canal.

The program includes the following figures for the main canal and separately for the laterals:
- Initial water flow;
- Final water flow;
- Initial sediment concentration;
- Final sediment concentration;
- Initial and final bed level;
- Accumulated sediment deposition.

6.3 CONCLUSIONS

The mathematical model will predict the water flow and sediment transport together with the variation in bottom level of the canal and is based on an uncoupled solution of the water flow and sediment transport equations. The model can be used for simulating the sediment entrainment in an irrigation network under changing flow conditions and sediment characteristics during the whole irrigation season and over one or more years. The model can be used for evaluating the effects of the interrelation between irrigation practice (operation and maintenance) and sediment deposition. The direct effect of irrigation practices on the sediment deposition or erosion of earlier deposited sediment may include changes in discharge, changes in sediment load, flow control structures, controlled entrainment, operation and maintenance activities, diverted sediment load to the farmlands, etc. Sediment deposition in the canal reaches may affect the following hydraulic aspects, such as water level variation (overtopping of canals), water distribution at outlets and flow control structures.

The theoretical background and assumptions made in the model has been comprehensively analysed by Paudel (2010) for upstream controlled systems and confirmed by Munir (2011) for downstream controlled irrigation systems. The main conclusions can be summarized as follows:
- *comparison* of SETRIC results with other existing flow and sediment transport models, namely DUFLOW (IHE, 1998; Clemmens et al., 1993) for the hydraulic computations and with the SOBEK

river model (DHL, 1994) for the sediment transport simulations showed fairly good proximity between the predicted values of SETRIC and DUFLOW and the SOBEK river model.

− *model verification*: the hydraulic and sediment transport predictions of the model are comparable with other models that are being used for research and design. The model predictions are in line with the natural process of morphological changes due to the change in water flows and sediment loads;

− *model validation*: the model has been validated using measured field data as input variables and comparing the predicted morphological changes with measured values. The predictions of the model have been found to be in line with the observed deposition patterns in an irrigation canal;

− *application to field data*: for the simulation of the sediment process the model has used as input variable measured data from a secondary irrigation canal, namely measured water flow, sediment concentration and water delivery schedules of two irrigation seasons. The deposition values predicted by the model ranged between 52% and 68% of the measured quantity during the two seasons;

− *general use of the model*: the model can predict the sediment transport behaviour of an irrigation canal for specific water requirements and delivery schedules. The model predicts the amount of sediment that will be deposited or eroded and determines the required capacity or efficiency of the sediment removal facilities and how they should be operated. The model can also assist in the selection of an improved operation plan that will result in minimal deposition in the canal systems that have been designed without considering any sediment transport criterion;

− *downstream control*: the SETRIC model has been applied to an automatically downstream controlled irrigation canal in Pakistan to investigate the sediment transport and to verify the model's application for downstream-controlled irrigation canals (Munir, 2011). Four different flow conditions (design flow, existing flow, and 50% of the design flow and canal operations under Crop Based Irrigation Operations (CBIO)) have been analysed for the hydraulic and sediment transport simulations in two irrigation schemes in Pakistan. The model was calibrated with field measurements in terms of canal roughness and the canal roughness was adapted to meet the measured and simulated water levels and for the sediment transport calibration from the simulations followed a multiplication factor for the equilibrium sediment transport formula for the specific conditions in Pakistan;

− *simulation of the sediment transport for downstream control*: along with the hydraulic computations the model proved to be able to simulate the sediment transport in downstream-controlled canals. The sediment transport has been simulated for several canal operation scenarios i.e. the design discharge, existing flow conditions and the CBIO. The total deposition as predicted by SETRIC was quite close to the sediment deposition by the SIC Model. Also the predicted sediment deposition for the existing water and sediment inflow were fairly close to the measured sediment deposition. The results strengthen the model's calibration and validation by showing a close match between the simulated deposition values and the measured values.

CHAPTER 7

The Sediment Transport Model
and its Applications

7.1 INTRODUCTION

The amount of water and sediment that enters into an irrigation canal will vary during the growing season and, moreover, throughout the entire life of an irrigation system. Variations in crop water requirement, water supply, size of the irrigated area, planned cropping pattern and sediment concentration frequently occur during the lifetime of irrigation systems. The design of canals and control structures incorporates a certain degree of flexibility in the delivery of different irrigation flows at fixed or variable supply levels. This design approach also assumes that the conveyance of the incoming sediment is, for the given design conditions, in a state of equilibrium. Once the flow conditions diverge from the design values, the flow velocity and thus the capacity to transport the sediment load will vary in time and space along the irrigation network. In these circumstances the initial assumptions related to the conveyance of the sediment load in or below equilibrium conditions are no longer valid for these changed flow conditions. Due to these changes, the sediment transport in irrigation canals will essentially be under non-equilibrium conditions. Therefore, the transport will strongly depend on the variation of the initial design flow and the changes in the incoming sediment load during the irrigation season and the lifetime of the canal. For this reason sediment transport should be viewed in a more general context, which should take into account the time and place of the varying operation requirements of the irrigation system.

The sediment transport model SETRIC offers the possibility to predict the sediment deposition and erosion in time and space, and for particular flow conditions and incoming sediment loads. This chapter will present some examples of sediment transport modelling in irrigation canals with the aim of showing the possible applications of this original model and of improving the understanding of sediment transport processes for situations commonly encountered in irrigation systems.

Sediment that enters a canal network can either be transported without any deposition along the canal or the sediment concentration can adapt itself from a non-equilibrium condition to the equilibrium sediment transport capacity of the canal. The adjustment towards the equilibrium transport capacity is assumed to follow Galappatti's depth-integrated model. A mass sediment balance in each canal reach will result in either net deposition or net entrainment. When the incoming sediment load is larger than the sediment transport capacity of the reach, deposition will occur. When the incoming sediment load is less than the transport capacity two possibilities can be identified. In the first case entrainment of previously deposited sediment occurs until the sediment transport is fully adapted to the equilibrium transport capacity. In the second case, no entrainment takes place and the incoming sediment load is conveyed without any change.

In the following examples the sediment deposition over a certain irrigation period will be simulated for an irrigation canal, for which the geometrical and hydraulic design data and the incoming sediment characteristics will be described in detail. The SETRIC model will be used to evaluate the effects of the following cases on the sediment behaviour:

1. Changes in the design discharge;
2. Changes in the incoming sediment load (concentration and diameter);
3. Controlled sediment deposition by deepening or widening of a canal section;
4. Overflow and undershot structures for upstream flow control;
5. Maintenance activities in a canal network;
6. Management activities in a canal network;
7. Effect of the design of control structures on the hydraulic behaviour and sediment transport.

7.2 DESCRIPTION OF THE HYDRAULIC AND SEDIMENT CHARACTERISTICS FOR THE GIVEN EXAMPLES

The examples in this chapter will start with an analysis of the effect of changes in the incoming discharge (Case 1) and incoming sediment load (Case 2) on the sediment deposition in an irrigation canal. Case 1 will extensively describe the differences in sediment deposition for the three sediment predictors that are available in SETRIC. The behaviour of the sediment for the cases 1 to 4 will be analysed for a single, straight irrigation canal with a trapezoidal cross section and a constant water depth at the downstream end (set point). The main hydraulic conditions and sediment characteristics during the simulation include:

• the discharge is constant during the simulation period;
• the water level at the downstream end of the main canal is kept at a set point that is determined by the normal water depth for the design flow and without sediment deposition in the canal; for other discharges the flow will be gradually varied;
• the downstream water level is managed by an underflow structure for cases 1, 2 and 3;

- the sediment size is kept constant for cases 1 to 4 and the incoming sediment concentration (in ppm) is kept constant for cases 3 and 4 during the whole simulation period;
- only sediment previously deposited on the canal bottom can be entrained during a later period of the simulation;
- the side slopes of the canal are stable;
- the initial roughness will be characterized by one single roughness along the whole wetted perimeter of the canal (Nikuradse, de Chézy or Manning);
- variations of the roughness might occur over space and time due to the occurrence of bed forms and the obstruction by weed growth;
- the sediment transport capacity at a certain place and time is given by the equilibrium concentration expressed in ppm by weight (parts per million);
- the time step and length step are based on a Courant number smaller than 1; for normal conditions a time step of 1–3 hours has been used and the length step is smaller than $L_{99\%} = 4.61L_A$ (see Section 5.3.5).

The geometrical, hydraulic and sediment characteristics of the irrigation canal are:

- length (L) $= 10\,000$ m
- bottom width (B) $= 10$ m
- side slope (m) $= 2$
- bottom slope (S_o) $= 0.00008$ (0.08 m/km)
- equivalent roughness (k_s) $= 0.018$ m (Nikuradse)
- bottom level (begin) $= +10.00$ m
- bottom level (end) $= +9.20$ m
- design discharge (Q) $= 25.75$ m^3/s
- normal water depth $= 2.38$ m
- water level at the downstream end $= +11.58$ m
- sediment size (d_{50}) $= 0.15$ mm
- incoming sediment concentration $= 100$ ppm (only for Case 1)
- simulation period $= 90$ days
- flow control structure $=$ undershot type
- flow condition $=$ uniform flow for the design discharge

7.3 CASE 1 CHANGES IN THE DISCHARGE AT THE HEADWORKS

A main objective of the design and operation of irrigation canals is to find the correct flow conditions for which the design water requirements are met for minimum or no sediment deposition. The most important reason for sediment deposition in the irrigation canals is a reduction in the water supply, caused either by a drop in crop water requirement or in water availability or by a variation in cropping pattern and the need to deliver water at the supply level (set point) to the command area.

For discharges smaller than the design discharge the flow in any canal reach might be either uniform flow (normal state) or a backwater flow. In this example (Case 1) the water level at the downstream end is kept constant at 2.38 m for discharges smaller than the design value and the flow profile will be a backwater curve. The sediment deposition that will occur in the canal is due to the combined effect of a reduction in discharge and the backwater effect. The reduction in discharge only will result in a smaller velocity, water depth, hydraulic radius and an increase in roughness; the backwater profile will result in a reduction in velocity and an increase in water depth and hydraulic radius and in a decrease in roughness. At the end of this example the two effects will be compared separately to show the effect of a reduction in discharge and of a backwater profile.

First, the effect of a reduction in water supply on the sediment transport and the deposition will be presented for a constant water level at the downstream canal end (see Figure 7.1). The reduction of the discharge is specified in terms of a relative discharge (Rd), which reads:

$$\mathrm{Rd} = \frac{\text{actual discharge}}{\text{design discharge}} \; (Q = 25.75 \, \mathrm{m}^3/\mathrm{s}) \qquad (7.1)$$

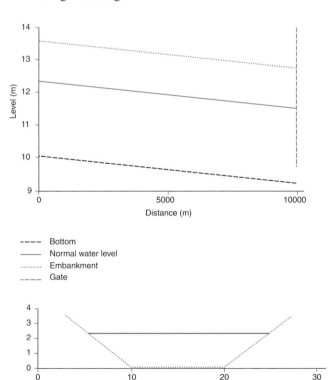

Figure 7.1. Longitudinal profile and cross section of the irrigation canal in Case 1.

The canal design has been based on the criteria of the rational method (see Chapter 4) and for uniform flow. From an irrigation and sediment transport point of view the design is not the most optimal solution, but otherwise these imperfections form an excellent basis to show the various aspects of SETRIC and how the programme can help designers and managers to improve their design or their operational activities. A better design should result in a canal that has a steeper bottom slope, a greater bottom width and a smaller water depth. The final design values will greatly depend on the sediment predictor that is appropriate for the specific design circumstances.

The design discharge (relative discharge $Rd = 1$) transports the sediment in equilibrium conditions according to Brownlie's predictor; the equilibrium transport capacity (108 ppm) is slightly larger than the incoming sediment load (100 ppm) and remains constant during the whole simulation period. The normal water depth is 2.38 m, the roughness C is $62\ m^{1/2}/s$ and the maintenance is ideal. The SETRIC results show that there is no deposition or erosion in the whole canal. Figure 7.2 shows the equilibrium and the actual concentration for the relative discharges $Rd = 1$ at the end of the simulation period of 90 days.

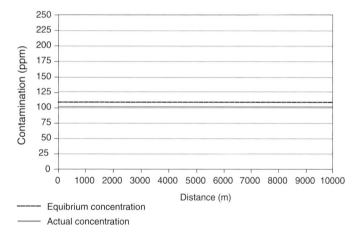

Figure 7.2. Equilibrium and actual concentration according to Brownlie's predictor at the end of the simulation period for Case 1 and $Q = 25.75\ m^3/s$.

Once the discharge decreases a gradually varied flow will develop due to the constant water level (set point) at the downstream canal end. The sediment transport capacity (equilibrium concentration) will also show a gradual decrease in the downstream direction due to the backwater caused by the downstream structure, where the actual water depth is larger than the normal water depth for flows smaller than the design discharge ($Rd < 1$) (see Table 7.1). For these discharges the equilibrium concentration is smaller than the incoming sediment load and this will result in a deposition that propagates from the upstream end in the downstream direction.

Table 7.1. Equilibrium concentrations and total amount of deposited sediment as a function of the relative discharge according to the Brownlie predictor.

Discharge (m³/s)	25.75	23.18	20.6	19.05
Rd	1	0.9	0.8	0.74
Equilibrium concentration (ppm)	108	91	77	70
Total volume of sedimentation (m³)	0	3 207	6 090	6 810

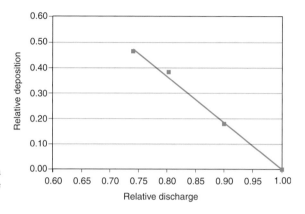

Figure 7.3. Relative sediment deposition at the end of the simulation period for Case 1 as a function of the relative discharge (according to Brownlie).

In Case 1, the incoming sediment concentration (100 ppm) remains constant for all the Rds and for the whole simulation period, but the total volume of sediment entering the system decreases for smaller Rds. Smaller discharges give smaller velocities that may result in more deposition of the sediment. Due to the fact that the total volume of incoming sediment during the simulation period depends on the relative discharge (Rd), a relative deposition will be introduced to compare the sediment depositions for various Rds. This relative deposition is the ratio of the sediment deposited in the entire canal and the total incoming sediment load at the head, both during the simulation period of 90 days:

$$\text{Relative deposition} = \frac{\text{total deposition in canal}}{\text{total volume of incoming sediment load}} \qquad (7.2)$$

Case 1 has also been used to compare the three sediment predictors that are included in the SETRIC programme, namely Ackers-White, Brownlie and Engelund-Hansen. For the above mentioned design characteristics the sediment deposition in the canal has been evaluated for the three predictors. The equilibrium concentration according to Ackers-White and Engelund-Hansen for the design discharge is lower than for Brownlie, which means that sediment deposition will already occur for these predictors at the design discharge. For smaller discharges the total volume of deposition will increase for both predictors in the same way as for the Brownlie predictor (Figure 7.3), but the total amount of sediment deposited will be larger (see Table 7.2).

Table 7.2. Comparison of the total amount of sedimentation and the relative sedimentation for the three predictors as a function of the discharge and the relative discharge.

	Discharge m^3/s	25.75	23.175	20.66	19.055
	Rd discharge	1	0.9	0.8	0.74
	Inflow sediment (m^3)	20023	18021	16129	14817
Brownlie	Volume sedimentation (m^3)	0	3207	6090	6810
	Rd sedimentation	0.00	0.18	0.38	0.46
Acker-White	Volume sedimentation (m^3)	7445	8313	8664	8414
	Rd sedimentation	0.37	0.46	0.54	0.57
Engelund-Hansen	Volume sedimentation (m^3)	5498	7990	9043	8792
	Rd sedimentation	0.27	0.44	0.56	0.59

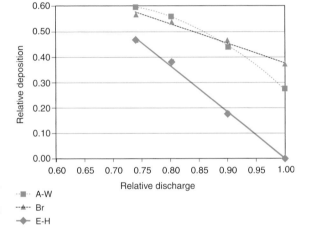

Figure 7.4. Relative sediment deposition as a function of the relative discharge at the end of the simulation period for Case 1 according to the Ackers-White, Brownlie and Engelund-Hansen predictors respectively.

Figure 7.4 shows the relative deposition as a function of the relative discharge Rd of 1, 0.90, 0.80 and 0.74, respectively. At the end of the simulation period the difference between the relative deposition for Brownlie and for the other two predictors is smaller for Rd = 0.74 than for the larger Rds. The comparison of the sedimentation for the Ackers-White and Engelund-Hansen predictors has been based on their equilibrium concentration for the design discharge of 25.75 m^3/s, which means that the relative deposition is larger than zero for a relative discharge of 1, specifically in Case 1. The comparison has not been extended to the larger discharges for which the incoming concentration (according to the Engelund-Hansen predictor) are the same for uniform flow at that discharge. The water depth, discharge and related velocity will all be different for the three predictors and a comparison will be more demanding.

Figure 7.5 shows the relative deposition in the canal for discharges smaller than the design discharge and for the cases where the flow is either uniform or a backwater flow. For the discharges smaller than the design discharge, the water depth in the whole canal at the start of the simulation is the uniform flow depth. The sediment deposition that will occur due to the effect of a reduction in discharge (Rd < 1) only is due to the fact that the flow velocity, water depth and hydraulic radius will be smaller and the roughness will be larger than the design values. The resulting sedimentation for this case is smaller than for a backwater flow.

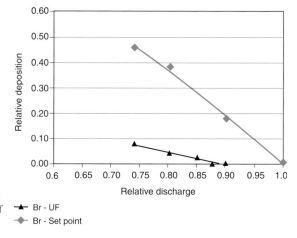

Figure 7.5. Relative sediment deposition (according to Brownlie) at the end of the simulation period for the case where the flow is uniform and for the cases where a set point is maintained, both as a function of the relative discharge.

7.4 CASE 2 CHANGES IN THE INCOMING SEDIMENT LOAD

The sediment load entering an irrigation system depends on the specific sediment conditions of the river and it is evident that the sediment concentration will vary during the irrigation season and even more during the lifetime of the system. Case 2 will evaluate the changes in incoming sediment that will include the variation in concentration as well as in the sediment size. For Case 2, the changes will be related either to the equilibrium concentration (108 ppm according to the Brownlie predictor) or to the median sediment size (0.15 mm) as assumed for the design of the irrigation canal (see Case 1).

To evaluate the sediment deposition behaviour due to changes in the concentration and/or mean size (d_{50}), the simulation period for all situations will be 90 days. Variations in the sediment characteristics are expressed in terms of either the ratio between the actual incoming sediment load and the sediment transport capacity of the canal (equilibrium concentration) or the ratio between the actual median sediment size and the design value of the median diameter (0.15 mm):

$$\text{Relative sediment load} = \frac{\text{actual sediment load}}{\text{equilibrium sediment load}\,(108\,\text{ppm})} \qquad (7.3)$$

$$\text{Relative sediment size} = \frac{\text{actual median sediment}}{\text{design median sediment size}\,(d_{50} = 0.15\,\text{mm})}$$
$$(7.4)$$

The changes in the sediment concentrations and diameter are kept within the limitations of Galappatti's model as presented by Ghimire (2003), Ribberink (1986), Wang et al. (1986) and Paudel (2010). Their research results are discussed in Chapter 5 and include two specific limitations: firstly, concerning the deviation of the mean concentration from the mean equilibrium concentration and secondly, the fall velocity of the sediment w_s/u_* should be much smaller than unity.

In view of the fact that the total incoming sediment load during the simulation period is different for each relative sediment load, a relative deposition value will be used to describe the deposition in the entire canal. The relative deposition is expressed in terms of the ratio of the total sediment deposited and the total amount of sediment entering the canal, both during the simulation period. It is expressed as:

$$\text{Relative deposition} = \frac{\text{total deposition in canal}}{\text{total volume of incoming sediment load}} \qquad (7.5)$$

The deposition in an irrigation canal depends on many factors; one is formed by the characteristics of the incoming sediment. Any deviation from the equilibrium concentration at a certain place and time will result in a relatively larger sediment load and a greater deposition in the entire canal. Table 7.3 shows the total amount of sediment deposited as a function of the relative sediment load after 90 days for the design flow of 25.75 m³/s and according to Brownlie's predictor.

Table 7.3. Total amount of sediment deposition as a function of the relative discharge according to Brownlie's sediment predictor.

Relative sediment load	1.0	1.25	1.5	2.0
Total volume of sedimentation (m³)	0	1889	5062	11392

Figure 7.6 shows the behaviour of the actual and equilibrium concentration as a function of time and place along the 10-km long canal. The incoming sediment load is 200 ppm and the actual and equilibrium concentrations are derived from Brownlie's predictor method. The decrease after the start of the simulation is very sharp, namely from 200 ppm to an equilibrium value of about 108 ppm. The decrease becomes gentler with time and moreover it will move in a downstream direction. The figure shows the concentration immediately after the start of the simulation and after 30, 60 and 90 days. After 90 days the actual concentration is about 112 ppm at a location 3 600 m from the intake. Along the whole canal the actual concentration is always slightly larger than the equilibrium concentration at any point.

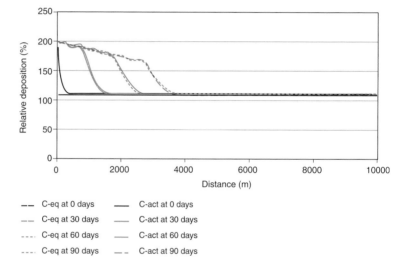

Figure 7.6. Changes in the actual and equilibrium concentration as a function of place and time according to Brownlie's predictor.

— — C-eq at 0 days	—— C-act at 0 days
— — C-eq at 30 days	—— C-act at 30 days
- - - C-eq at 60 days	—— C-act at 60 days
- - - C-eq at 90 days	— — C-act at 90 days

Figure 7.7 shows the behaviour of the sediment deposition as a function of time and place along the 10-km long canal. The incoming sediment load is 200 ppm and due to the fact that the actual concentration is higher than the equilibrium concentration the sediment will deposit on the bottom, starting from the intake. Next the deposition will move in a downstream direction and at the same time the height of the deposition will increase in the area where sediment has already been deposited. The slope of the front of the deposition will also become gentler when it moves in a downstream direction. The observations for the changes in concentration and movement of the deposition are comparable for the three sediment transport predictors. Figures 7.5 and 7.6 are based on Brownlie's predictor and are given as examples.

The sediment deposition in the canal as a function of a larger sediment diameter is presented in Table 7.4 for the Brownlie predictor. The table shows the total volume of sediment deposition after 90 days

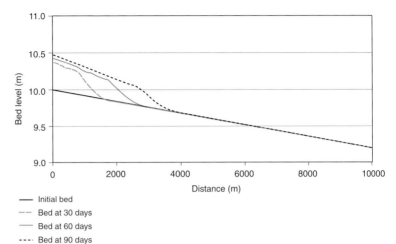

Figure 7.7. Changes in the bed level from the original bed level after 30, 60 and 90 days according to Brownlie's predictor.

Table 7.4. Total amount of sediment deposition as a function of the relative sediment diameter according to Brownlie's sediment predictor.

Sediment diameter (mm)	0.150	0.200	0.250
Relative sediment diameter	1	1.33	1.66
Total volume of sedimentation (m^3)	0	2 292	4 229

and for a design flow of 25.75 m^3/s. Deviation from the design diameter ($d = 0.150$ mm) for the equilibrium concentration according to Brownlie (108 ppm) results in relatively more deposition in the entire canal (Figure 7.8).

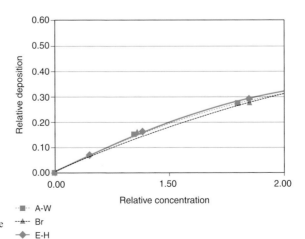

Figure 7.8. Relative sediment deposition after 90 days as a function of the variation in the relative concentration and for the three predictors.

The same conditions are also used to compare all the three sediment predictors, namely Ackers-White, Brownlie and Engelund-Hansen. For the same design characteristics as presented in Case 1, the sedimentation in the canal has been evaluated for the three sediment transport predictors. The equilibrium concentration according to the Ackers-White and Engelund-Hansen predictors for the design discharge and a sediment diameter of 0.15 mm (see Case 1) are smaller than for Brownlie, namely 41 and 56 ppm. This means that sediment deposition according to these predictors will already occur for the design discharge of 25.75 m^3/s and an incoming sediment concentration of 100 ppm. The effect of a larger sediment diameter on the sediment deposition is presented in Figure 7.9. The figure shows that the changes in the relative sediment size have an evident impact on the deposition and that the deposition will increase in the same way for all three predictors.

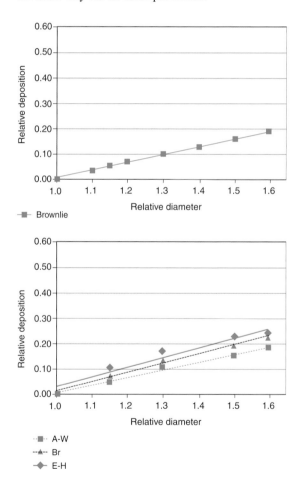

Figure 7.9. Relative sediment deposition after 90 days as a function of the variation in relative median sediment size for the three predictors (Q = 25.75 m^3/s and c = 100 ppm).

7.5 CASE 3 CONTROLLED SEDIMENT DEPOSITION

Uncontrolled sediment transport and deposition in irrigation canals may result after some time in clogging (obstruction) of tertiary turnouts, reduction of the canal conveyance capacity, large variation in water levels, unpredictability of the relationship between water depth and discharge for regulating structures, and high costs for canal desilting works. Once the sediment enters a canal network it can be conveyed by the canals to the farm plots or it can be removed from the water, for example, either by deposition along the entire canal or by deposition in a sediment trap within some of the canal reaches where it can be periodically removed at minimum cost. Controlled deposition can be obtained by the design of a sediment trap in the network, for example by deepening or widening one or more canal reaches. This solution can be an attractive alternative for controlling the deposition, but a careful selection of the location of the sediment trap during the design phase needs careful consideration, which needs a detailed elaboration in view of the future operational and maintenance costs.

Case 3 will present two scenarios to control the deposition of sediment and will show how a major part of the sediment load can be deposited in some adapted canal reaches. A change in a canal cross section may reduce the transport capacity (equilibrium concentration) in such a way that it becomes smaller than the actual sediment load and part of the sediment will be deposited there. In Case 3 the canal from Case 1 receives a sediment load of 150 ppm at the headworks and is not able to transport the sediments to the fields during the irrigation season. Therefore, when no special measures are taken to control the deposition, the latter will occur along the entire canal length. Case 3 will analyse two scenarios to control the sediment deposition in the head reach by reducing the transport capacity in that reach only.

The head of the irrigation canal will be converted into a settling basin that can be described as:
1. *Scenario 1*: deepening of the canal by 0.50 m for the first 1 000 m;
2. *Scenario 2*: widening of the bottom width from 10 m to 14 m for the first 1 000 m.

No other alternatives will be evaluated for a further optimization of costs and/or sediment deposition. The previously mentioned data (see Case 1) will also be used for the two scenarios in Case 3. For Scenario 1 the original wetted area (35.13 m^2) will increase by 4.5 m^2 and for Scenario 2 the increase will be 9.52 m^2, both for the same water level in that reach. The total amount of excavation required for Scenario 2 will be larger than 9.52 m^2 and depends on the freeboard, which might be 0.5 times the design water depth or more.

The simulation results of the two scenarios will be compared with the deposition in the canal without any provision to trap the sediment over the first 1 000 m (Case 1 with a sediment inflow of 150 ppm). Scenario 1 (deepening) holds less sediment than Scenario 2 (widening $B = 14$ m) for the same flow and sediment conditions of Case 1. After 90 days Scenario 1 has accumulated much more sediment than the canal without any sediment control over the first 1 000 m. The controlled deposition by the two scenarios is especially effective during the first part of the simulation period (about 60 days).

Figure 7.10 presents the sediment depositions over the first 1 000 m of the canal with and without controlled deposition (scenarios 1 and 2). This figure also gives the deposition if the canal is widened to 12 m (the same increase in the wetted area as the deepening). Table 7.5 gives the volumes of deposited sediment after 90 days together with the equilibrium concentrations near the headworks at the start and at the end of the simulation period. The deepening and the widening ($B = 12$ m) which have the same wetted area at the start of the simulation show comparable

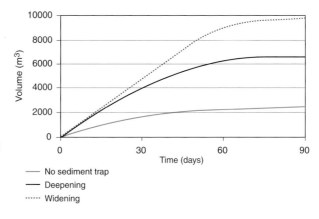

Figure 7.10. Sediment deposition in the irrigation canal with and without controlled deposition (deepening and widening of the bottom over the first 1 000 m).

Table 7.5. Total amount of sediment deposition over the whole canal length and the equilibrium concentration for the cases with and without measures to control the deposition over the first 1 000 m of the canal ($Q = 25.75$ m³/s and $c = 150$ ppm).

	No control	Deepening	Widening $B = 12$ m	Widening $B = 14$ m
Total volume of sedimentation (m³) after 90 days	5 063	8 180	7 622	10 708
Equilibrium concentration (ppm) for $x = 0$ and $t = 0$ day	108	56	65	37
Equilibrium concentration (ppm) for $x = 0$ and $t = 90$ days	147	148	147	146

Table 7.6. Total amount of sediment deposition over the first 1 000 m of the canal after 0, 30, 60 and 90 days.

Deposition over 1 000 m	Days	No control	Deepening	Widening $B = 12$ m	Widening $B = 14$ m
Volume of sedimentation (m³)	0	0	0	0	0
Volume of sedimentation (m³)	30	1 639	4 032	3 645	4 890
Volume of sedimentation (m³)	60	2 225	6 272	5 632	8 998
Volume of sedimentation (m³)	90	2 499	6 605	5 964	9 826

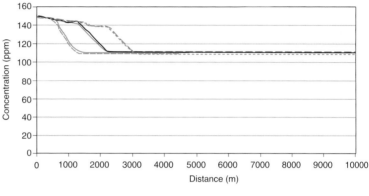

Figure 7.11. Changes in the actual and equilibrium concentration after 30, 60 and 90 days for a canal with no sedimentation control.

-- C-eq at 30 days   — C-act at 30 days

---- C-eq at 60 days   — C-act at 60 days

---- C-eq at 90 days   -- C-act at 90 days

behaviour in sediment deposition and changes in concentration (see Table 7.6). Widening of the canal ($B = 14$ m) leads in this example to the smallest equilibrium concentration. However after some time the equilibrium concentration for the two scenarios will be in line with the concentration of the canal without control measures (see Figures 7.11, 7.12 and 7.13).

The simulation results of the deposition due to the deepening of the canal are shown in Figure 7.12, which gives the actual sediment load and the equilibrium concentration for Scenario 1 (deepening) after 30, 60 and 90 days. When the sediment trap is almost empty the equilibrium concentration is very low, after 30 days the trap is partly filled and the equilibrium concentration increases; after 60 days the value is the same as in the situation without measures to control the sediment deposition (Case 1). The sediment trap (4 500 m³) will be filled in about 30 days, and after that the sediment will be deposited on the modified canal bottom in the same way as in the canal without control measures.

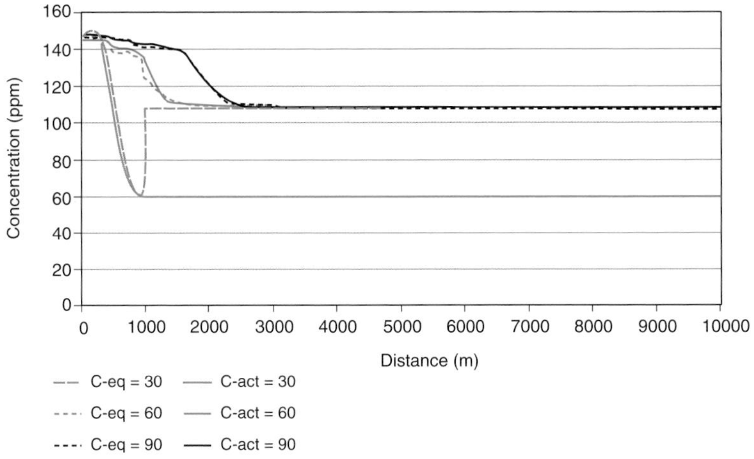

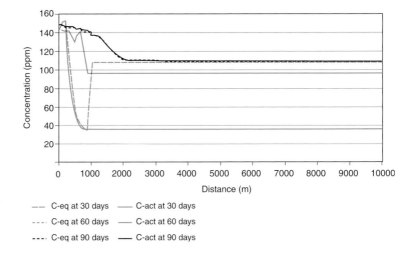

Figure 7.12. Changes in the actual and equilibrium concentration after 30, 60 and 90 days for a canal with sedimentation control (deepening).

Figure 7.13. Changes in the actual and equilibrium concentration after 30, 60 and 90 days for a canal with sedimentation control (widening $B = 14$ m).

The sediment trap gives small equilibrium concentrations and high deposition volumes during the simulation period. At the end of the simulation (90 days) the equilibrium concentration rises to the actual concentration and the point at which the sediment trap has to be cleaned becomes critical.

The simulation results for the deposition due to the widening of the first canal reach (Scenario 2 with $B = 14$ m) are shown in Figure 7.13. The behaviour of the actual and equilibrium concentration is rather similar to the one previously described (Scenario 1). The widening shows more deposition than the deepening of the canal. The table shows that the amount of sediment deposition per day for the canal without control measures is smaller than for the canal with one of the two control measures (scenarios 1 and 2). The amount of sediment deposition per day for the deepening scenario after 60 days is almost in line with the amount for the canal without measures, while the sediment deposition for the widening scenario ($B = 14$ m) at the end of the simulation period is still two times larger than for the deepening scenario.

7.6 CASE 4 FLOW CONTROL STRUCTURES

One of the main design criteria for irrigation canals is the command level, which states that the canals should be able to deliver irrigation water at the right elevation, at the correct time and at the right amount to the command area. In view of this criterion, flow control structures are needed to keep the water level at the command or target level for any discharge during the irrigation season. Two cases with a different type of control structure will be considered, namely:
– a structure with undershot flow (Case 4.1);
– an overflow structure (Case 4.2).

When an overflow structure has been correctly designed the flow velocity will not decrease for the design flow and no deposition will occur in front of the weir. If the overflow structure has not been correctly designed deposition will follow due to the backwater effect (reduction in velocity) that is created by the hydraulically faulty weir design. This aspect will be discussed in the last example (Case 7) that will show the difference between a hydraulically well designed and a hydraulically poorly designed flow control structure for fluctuating discharges. The example is useful to analyse the correct design of a flow control structure for a given water flow and sediment characteristics.

Case 4 will evaluate the ability of overflow and undershot structures to convey sediments; operational as well as other aspects that might influence the selection of a structure will not be discussed. Bed load and suspended load are freely conveyed by all structures with undershot flow; overflow structures will also convey the suspended load, but any bed load will be trapped. For the comparison both control structures are placed 2 500 m downstream from the headworks and they will maintain a uniform flow for the design discharge (25.75 m^3/s). Downstream of both structures is a bottom drop of 0.40 m to guarantee modular flow over the structures, which means that the downstream water level will not influence the discharge and hence the water level upstream of the structure.

The behaviour of the two structures in respect of the sedimentation will be compared on the basis of the changes in sediment concentration and in bed level. The major changes in concentration for both structures occur directly downstream of the headworks. The incoming concentration is 150 ppm and the behaviour of the actual and equilibrium concentration as a function of time and place along the 10 km long canal is comparable with the actual and equilibrium concentration of Case 2 (Section 7.4). The actual and equilibrium concentrations are derived from the Ackers-White predictor and the decrease in concentration after the start of the simulation is very sharp, from 150 ppm to an equilibrium value of approximately 56 ppm. The decrease becomes gentler with time and moves in a downstream direction. Figure 7.14 presents both the concentrations immediately after the start of the simulation and after 30, 60 and 90 days. After 90 days the actual concentration is about 150 ppm near the gate and drops gently to 56 ppm downstream of the headworks. Along the whole canal the actual concentration is always slightly larger than the equilibrium concentration. Figure 7.15 presents the changes in bottom level in the canal, mainly upstream of the gate at 2 500 m. The sedimentation moves from the headworks in a downstream direction. After 90 days the increase in bed level at the headworks is about 0.40 m and decreases in the downstream direction towards the gate, where it is less than 0.30 m. At the gate location no sedimentation occurs, but the sediment has proceeded downstream of the gate where a layer of about 0.10 to 0.15 cm can be observed. The sedimentation will move farther downstream with time.

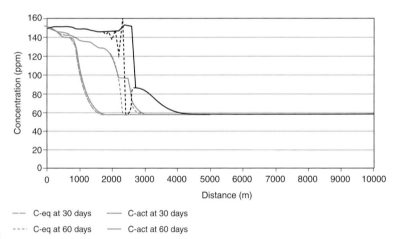

Figure 7.14. Changes in the concentrations in the canal with a gate at 2 500 m from the intake after 30, 60 and 90 days.

－－ C-eq at 30 days   —— C-act at 30 days
- - - C-eq at 60 days   —— C-act at 60 days
- - - C-eq at 90 days   —— C-act at 90 days

Figure 7.16 shows the concentrations immediately after the start of the simulation and after 30, 60 and 90 days for the case where a weir is placed 2 500 m from the headworks. The actual concentration behaves in the same way as the concentration for the situation with a gate at 2 500 m. The behaviour of the concentration changes when the sedimentation reaches the weir after approximately 60 days. The bed load will be trapped by the weir and the canal bottom level will increase upstream of the weir.

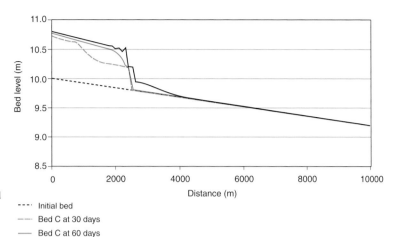

Figure 7.15. Changes in the bed level in the canal with a gate at 2 500 m from the intake after 30, 60 and 90 days according to Brownlie's predictor.

---- Initial bed
—-— Bed C at 30 days
—— Bed C at 60 days
—— Bed C at 90 days

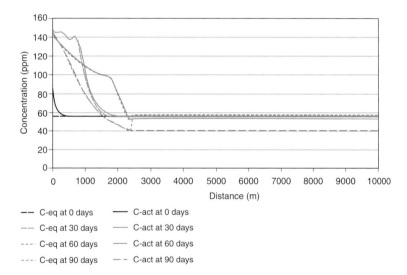

Figure 7.16. Changes in the concentration in the canal with a weir 2 500 m from the intake after 30, 60 and 90 days according to Brownlie's predictor.

—— C-eq at 0 days   —— C-act at 0 days
—— C-eq at 30 days   —— C-act at 30 days
---- C-eq at 60 days   —— C-act at 60 days
---- C-eq at 90 days   —— C-act at 90 days

This phenomenon is the most striking difference in behaviour between the gate and the weir. The gate will not trap any sediment inside the structure; the weir will trap the sediment in the upstream canal section until a new bottom slope has been developed for which the equilibrium concentration is in line with the incoming sediment concentration. Figure 7.16 clearly shows that the change in concentration after 90 days differs from Case 4.1 (Figure 7.14). The sedimentation upstream of the weir will continue until the bottom level is steep enough to transport the incoming sediment to the downstream canal section. Figure 7.16 also shows smaller equilibriums

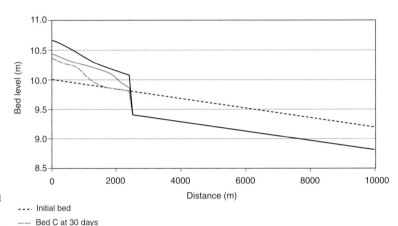

Figure 7.17. Changes in the bed
level in the canal with a weir at
2 500 m from the intake after 30,
60 and 90 days according to
Brownlie's predictor.

and actual concentrations upstream of the weir than for the gate, both after
90 days. The concentration even decreases in relation to the values after
30 and 60 days.

An explanation for the location of the gate and the weir at 2 500 m
from the headworks in this example is the fact that when these structures
are placed at the end of the 10 000 m long canal that the sediment for
both structures has only progressed up to 4 000 m after 90 days and no
difference in sedimentation behaviour between a weir or a gate could be
observed.

7.7 CASE 5 MAINTENANCE ACTIVITIES

The roughness of the bottom and side slopes will have a specific impact
on the sedimentation in an irrigation canal. SETRIC will calculate the
roughness on the canal bed and sides slopes separately; the roughness on
the bed is determined with the van Rijn (1984C) method, which includes
the flow conditions and the bed form and grain-related parameters. The
next step is the equivalent roughness coefficient (k_s) that takes into
account the sidewall effect (Mendez, 1998 and Paudel, 2010); the model
computes the total friction factor from the composite roughness of the
cross section. The roughness conditions will vary over time, especially
sedimentation will induce the development of bed forms and the total
equivalent friction factor is computed for every time step and for each
flow condition in each cross section.

Two aspects of the effect of the bed and side slope roughness on the sediment deposition as a result of maintenance activities will have to be considered:

a. When the canal is ideally maintained the roughness remains equivalent to the design condition; the roughness is the same as at the design state with the water level at the required set point for the design flow. For flows smaller than the design discharge a backwater condition will be created to maintain the set point.

b. The case that the canal is poorly maintained and the roughness will increase with time. After some time the water depth at the design discharge will increase above the design water level, which means that either less irrigation water flows through the canal for the same set point or that the water flow can only be conveyed for water levels higher than the set point.

In this example (Case 5) the set point will be adapted to the changing roughness to maintain the normal depth for poor maintenance conditions. This change in set point level is one of the solutions to cope with the changes in roughness. Depending on the actual management policy of an irrigation system other scenarios can also be analysed to evaluate the effect of the changing roughness on the water management and sediment transport. The model can be a unique tool to analyse the various scenarios to cope with specific maintenance activities in the canal network.

The roughness depends on the construction of the canal (i.e. lining, masonry, earthen canal) and on the maintenance during or after the irrigation season. This section will describe a simple example for the impact of the roughness on the sedimentation. In this case the canal from Case 1 will be simulated for an ideally maintained canal as well as for a poorly maintained canal. The cross section, longitudinal profile and the length are the same as in Case 1. For both cases the flow will be uniform and is controlled by a gate at the downstream end of the canal. Upstream of this gate the water depth is 2.45 m for the poorly maintained (C at approximately 60) and 2.35 m for the ideally maintained canal (C at approximately 64). The canal receives a sediment load of 150 ppm at the headworks and is not able to transport the sediment to the fields during the irrigation season. The equilibrium concentration is 59 ppm for the canal with ideal maintenance and 54 ppm for the canal with no maintenance, both according to the Ackers-White sediment predictor.

Figure 7.18 shows the changes in concentration for the canal with ideal and with no maintenance. The equilibrium concentration in the canal with ideal maintenance is higher than for the canal with no maintenance; the progress of the changes in the canal with ideal maintenance is faster than in the canal with no maintenance. After 90 days the ideally maintained canal has accumulated 11 501 m^3 and the canal with no maintenance

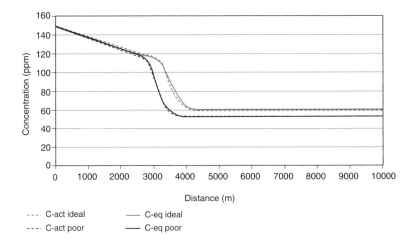

Figure 7.18. Changes in concentration after 90 days for an ideally and for a poorly maintained canal according to the Ackers-White predictor.

12 281 m^3 over the full canal length. Figure 7.19 presents the sediment depositions for both cases. The height of the sedimentation for the ideally maintained canal is lower and its length is slightly longer (about 400 m) than for the canal with no maintenance.

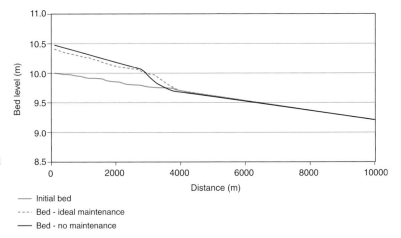

Figure 7.19. Changes in the bed level from the original bed level after 90 days for an ideally maintained and for a poorly maintained canal according to the Ackers-White predictor.

7.8 CASE 6 MANAGEMENT ACTIVITIES

Management of an irrigation system comprises all the necessary operation and maintenance activities to deliver water to the users at the correct right time, at the right level and with the right amount. Each irrigation scheme is operated in its own, often different way. The operation depends on various factors, such as the water availability, management of the water supply, scheduling of water delivery, control of water levels and discharges, water measurement, qualification of the operation personal, institutional limitations, field water requirements, water rights, evaluation and monitoring and maintenance activities. This large range of the numerous options and constraints makes it difficult to develop a general selection procedure for the most effective operation (Walker, 1993). Formulation of an operation plan for a specific irrigation system requires the evaluation of several scenarios in view of the reliability of the water delivery. One of the aspects related to the reliability is the risk of sediment deposition in the canal system. The simulation of several operation scenarios will support the evaluation of various effects of the operation on the sediment deposition in a canal network. In order to present a first assessment of these effects a canal network will be schematized and the main data will include the irrigation requirements, the geometrical and hydraulic data and incoming sediment characteristics.

The network consists of a 16-km long main canal that delivers water to a command area of 10 000 ha by 6 offtakes along and at the downstream end of the canal. The canal has six reaches, namely AB, BC, CD, DE, EF and FG. The first offtake at the downstream end of AB delivers water to a command area of 1 800 ha that is 3 km from the head of the main canal. The second offtake conveys water to a command area of 1 600 ha and is 6 km downstream of the headworks. The third reach, CD, conveys water to the offtake 3 that irrigates an area of 1 500 ha and is 10 km from the headworks. The offtake at 12 km from the headworks will irrigate 1 600 ha. At the end of the canal is a water level regulator where the water is conveyed to two areas, 1 800 and 1 700 ha in size respectively. All the offtakes are located upstream of water level regulators. Figure 7.20 shows the schematization of the network and the longitudinal profile of the main canal.

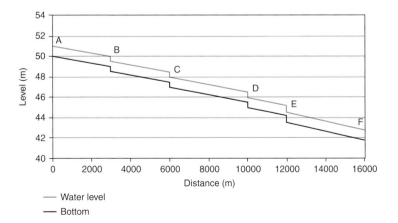

Figure 7.20. Longitudinal profile of the main canal ABCDEF.

Table 7.7. Main characteristics of the main irrigation canal.

Reach	AB	BC	CD	DE	EF	
Offtake number	1	2	3	4	5	6
Location from headworks (km)	3	6	10	12	16	16
Command area (ha)	1 800	1 600	1 500	1 600	1 800	1 700

Table 7.8. Geometrical data of the main irrigation canal ABCDEF.

Reach	Length L (m)	Width B (m)	Chézy's roughness coefficient	Side slope m (–)	Bottom slope S_o (10^{-3})	Normal water depth (m)
A–B	3 000	5.15	42	1.5	0.345	1.19
B–C	3 000	5.15	42	1.5	0.345	1.19
C–D	4 000	4.00	42	1.5	0.375	1.04
D–E	2 000	4.00	42	1.5	0.375	1.04
E–F	4 000	2.70	42	1.5	0.425	0.83

Table 7.9. Water needs during the irrigation season of the irrigated area.

	Irrigation period of 10 days during the irrigation season				
Period	1	2	3	4	5
l/s·ha	6	5	4	6	5

Table 7.7 shows the main characteristics and Table 7.8 gives the geometrical and hydraulic characteristics of the main canal.

The whole irrigation season is divided into 5 periods of 10 days. Table 7.9 gives the water requirement at the headworks for each period.

For the simulation of the effects of the operation plan on the sediment deposition in the main canal, a few scenarios are analysed. All of them comply with the water requirements and the water supply for the irrigated areas. The Ackers-White sediment transport predictor is used to compute the sediment transport capacity. In this hypothetical case the distribution efficiency of the main canal is assumed to be 100%. The incoming sediment load at the headworks during the whole irrigation season is characterized by a median diameter d_{50} of 0.15 mm. The incoming sediment concentration varies per period of 10 days.

The scenarios are:

- *Scenario 1 (continuous flow)*: a continuous flow to all offtakes during the whole season (see Figure 7.21);
- *Scenario 2 (rotational flow by 5 days)*: a rotational flow during a period of 10 days towards the groups A and B; each group receives water for alternate periods of 5 days.

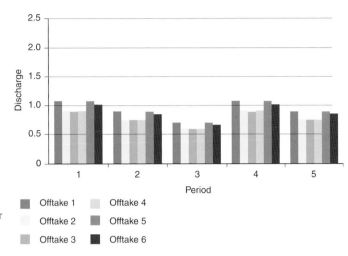

Figure 7.21. Flow to the offtakes along the main canal for continuous flow during the 5 periods of the growing season.

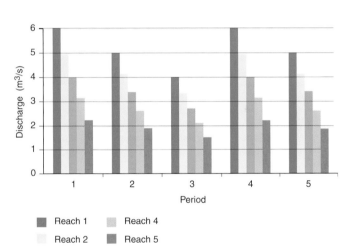

Figure 7.22. Flow to the offtakes along the main canal for rotational flow during the 5 periods of the growing season.

The 6 offtakes are grouped in two groups, namely Group A with the offtakes 1, 3 and 6 and Group B with the offtakes 2, 4 and 5 (see Figure 7.22). The flow in the reach AB is the same for the rotational and continuous flow during a period of 10 days.

The deposition during the irrigation season of 50 days is determined for the entire canal ABCDEF and for the two scenarios. The water level regulators are set every 5 days for the rotational flow in the main canal and every 10 days for continuous flow in the canal. The same applies to the offtakes that are set every 5 days for the rotational flow.

During the first period of 10 days the equilibrium sediment concentration is larger than the incoming, actual sediment load, in terms of time as well as space, for the rotational and for the continuous flow. In the second and third period the incoming sediment load is larger in a few canal reaches than the equilibrium concentration for the rotational flow, especially in the reach DE. In this reach sediment will be deposited with a maximum height of about 5 cm. In Period 4 the actual sediment load is smaller than the equilibrium concentration and the already deposited sediment during the periods 2 and 3 will be eroded and transported out of the canal. During Period 5 the equilibrium concentration is larger than the incoming load, but no sediment will be picked up due to the fact that the original bottom level was reached at the end of Period 4.

The sediment behaviour for continuous flow is comparable with that for the rotational flow during periods 1, 2 and 3. However, during periods 4 and 5 the incoming load is relatively larger than the equilibrium concentration and the deposition of sediment will continue during periods 4 and 5, resulting in about $700 \, \text{m}^3$ sediment being deposited in the canal.

Figure 7.23 shows the total volume of deposition in the canal for respectively continuous and rotational flow at the end of each period and for a sediment inflow that changes with the incoming discharge as given in Table 7.10.

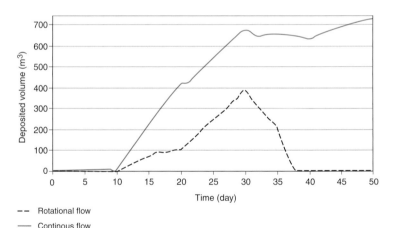

Figure 7.23. Sediment deposition in the canal network as a function of time for continuous and rotational flow according to Ackers-White.

-- Rotational flow
— Continous flow

Table 7.10. Incoming sediment concentration during the irrigation season.

	Irrigation period of 10 days				
Period	1	2	3	4	5
Sediment concentration (ppm)	200	300	200	300	200

The deposition in the canal has been analysed to see if the incoming sediment load is constant. Two cases have been evaluated, namely a constant inflow of 200 ppm and one of 300 ppm during the whole irrigation season. The simulation shows that a continuous flow gives more sedimentation than a rotational flow in this specific case. In view of the small differences the results are presented in Figure 7.24. The difference between a sediment concentration that changes with the water inflow and a constant sediment inflow is small for this canal system.

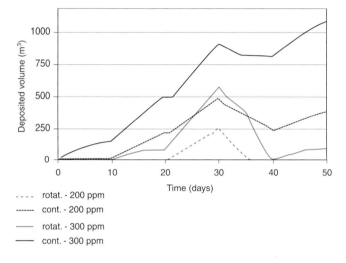

Figure 7.24. Sediment deposition in the canal network as a function of time for continuous and rotational flow and for two concentrations, according to Ackers-White.

- - - -   rotat. - 200 ppm

-------   cont. - 200 ppm

———   rotat. - 300 ppm

———   cont. - 300 ppm

A further investigation of the influence of maintenance activities on the behaviour of the sediment in the canal network with rotational flow resulted in minimal differences in sedimentation between an ideally maintained canal and a poorly maintained canal. The results are in line with the data discussed in Section 7.6.

7.9 CASE 7 EFFECT OF THE DESIGN OF CONTROL STRUCTURES ON THE HYDRAULIC BEHAVIOUR AND SEDIMENT TRANSPORT

Overflow and undershot structures have different effects on the hydraulic behaviour of an irrigation canal and on the sediment transport by the canal. The SETRIC model is a helpful tool to analyse the effect of well-designed overflow structures (without a gate) on the sediment transport process when there are fluctuations in the discharge. This case will present the difference between hydraulically well and hydraulically poorly designed flow control structures. This type of analysis will be helpful to finalize the design of a flow control structure for a given water flow and sediment characteristics.

This example will be based upon a canal with an upstream section of 2 500 m and a downstream section of 500 m. An overflow weir ($B = 10$, 12 and 14 m respectively) is situated at the end of the first section for a design discharge of 25.75 m³/s and an upstream water depth of 2.38 m. The inflow concentration is higher than the equilibrium concentration for this canal in order to compare the results with some of the previous examples (Case 2 and Case 4). This high concentration means that the sedimentation in the canal will be due to the combined effect of the higher concentration and the backwater profile. Even without a weir deposition will occur in the canal due to the high concentration. At the end of this section the effect of a case with a concentration lower than the equilibrium concentration at the design discharge will be discussed.

The design of the 3 broad-crested weirs, which will have the same water level as the canal for the design flow, will be evaluated (see Figure 7.25). Downstream of the weir is a canal section with a bottom level lower than the upstream bottom to guarantee a modular flow over the weir; the upstream water level and the discharge over the structure are independent of the downstream water level.

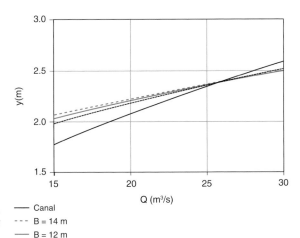

Figure 7.25. Water depth in the irrigation canal and upstream of a weir with a width B of 10, 12 and 14 m respectively.

For a wider weir the crest will be higher in order to maintain the set point, and for all discharges smaller than the design discharge the flow over all the weirs remains free and a backwater profile will develop in the upstream canal section. For smaller weir crests the backwater effect will be more pronounced than for wider ones. The width of the weir determines the rise in water level for smaller discharges and therefore the sediment deposition upstream of the weir.

The differences in sediment deposition, which are determined using the Ackers-White sediment predictor, are not very significant for the various widths. For the design discharge the relative sedimentation is about 0.6 and for the smallest width the relative sedimentation is about 0.85 (see Figure 7.26). This figure also gives the sediment deposition for a gate (without a weir) at the same location, which maintains the identical set point for the design discharge and smaller discharges as the design flow over the weirs.

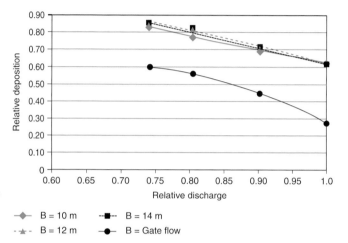

Figure 7.26. Relative deposition in the canal upstream of a weir with a width of 10, 12 and 14 m and of a gate respectively.

For the given sediment characteristics ($d_{50} = 0.15$ mm) and using the Ackers-White sediment predictor, the equilibrium concentration for the design discharge is approximately 56 ppm. When the incoming concentration is less than 56 ppm, only the structure will affect the sediment deposition in the canal. For the design discharge and for an incoming sediment concentration less than the equilibrium concentration there will be no deposition, which is valid for all weir widths that are designed for the same set point. For discharges smaller than the design discharge a backwater profile will develop that determines the sediment deposition in the canal.

The next example has an inflow concentration equal to the equilibrium concentration for the design discharge (55 ppm). The same incoming concentration is maintained for the simulation with smaller discharges to show the effect of mainly the backwater. The relative sediment deposition in the canal is for all weir widths smaller than in the previous example. The results for various weir widths are presented in Figure 7.27. Weirs with a small width result in relatively small relative deposition (15–25%) and the effect for wider weirs is slightly higher but still comparable (around 40%).

In addition the effect of the crest height on the sediment deposition for the design discharge has been evaluated for a weir width of 10 m.

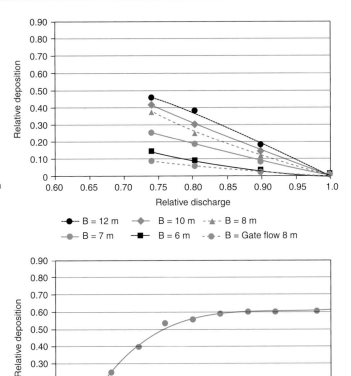

Figure 7.27. Relative deposition in the canal upstream of a weir with various widths and as a function of the relative discharge.

Figure 7.28. Relative deposition in the canal for various crest levels.

The incoming concentration equals the equilibrium concentration for the design crest height (55 ppm). The design crest level is +10.90 m above datum and no sedimentation in the canal is observed for this 10 m-wide weir. For higher weir crests the relative sediment deposition increases relatively sharply and above a certain level the relative deposition levels off (see Figure 7.28).

The relative deposition has a steep rise in the beginning and flattens off for higher crest levels. For the design discharge and an inflow concentration less than the equilibrium concentration, there is no deposition. For a crest height higher than the design level a backwater profile is created that will reduce the sediment transport capacity in the canal and deposition will take place. For higher crest levels more deposition will take place in the system and sediment flowing out of the system will decrease. At a certain crest level most of the inflowing sediment is deposited in the system, the concentration near the weir becomes low (1–2 ppm) and further increments of the crest level will not result in any quantifiable increase in sediment deposition. The analysis of a weir at 2 500 m from the intake shows that for an actual incoming concentration less than the equilibrium concentration there is no sedimentation in the canal.

The previous discussions show that for discharges smaller than the design discharge the equilibrium concentrations are smaller than the concentration for the design discharge (55 ppm). These smaller values have been used to evaluate only the effect of the structure. Therefore the equilibrium concentrations of the canal have been determined for various discharges without the influence of the weir and this concentration has been used as the incoming concentration for a simulation with the weir at the end of the first canal section. The effect of the structure on the deposition will be more pronounced when the sediment-carrying capacity of the canal decreases for smaller inflow discharges. The model simulations show some deposition in the canal, which started from the weir in an upstream direction. The values are very small, about 5–10% of the incoming sediment load, and the deposition height is also relatively small, but nevertheless observable. This phenomenon can also be observed in the previously presented examples, but it is very small and much smaller than the sedimentation due to the suspended load.

Figure 7.29 gives the sedimentation over the first 300 m in front of the weir for an incoming concentration smaller than the equilibrium concentration and for a discharge of 74% of the design discharge. The described tendency is very clear during the first days after the start of the simulation. For longer periods and for larger discharges the effect becomes smaller.

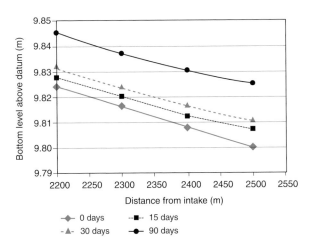

Figure 7.29. Sediment deposition in the canal section 300 m upstream of the weir as a function of time.

Figure 7.30 shows the sediment volumes deposited in the canal for a weir width of 10 m at the end of a canal section of 2 500 m for the various discharges (74, 80, 90, 100%) and for an incoming concentration equal to the equilibrium concentration at $t = 0$ and $x = 0$ (23, 32, 42 and 54 ppm respectively). The deposited volumes due to the structure are small

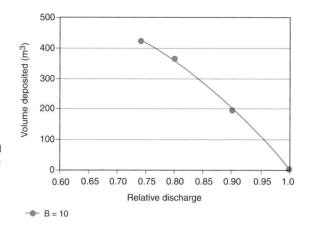

Figure 7.30. Volume deposited in the canal as a function of the relative discharge and for incoming concentrations equal to the specific equilibrium concentration at $t = 0$, $x = 0$.

(about 5–10%) of the total incoming sediment load. The weir has some effect on the deposition, but it is very small (mms or a few cms); the relative deposition as a function of the relative discharge is not given as it is small and almost constant with the relative discharge.

7.10 REVIEW

The previously discussed examples have shown the various aspects of sediment transport in irrigation canals and network. The dynamic behaviour of the sediment transport process with its various variables that are a function of place and of time is presented in the numerical and graphical results of the SETRIC model. To obtain the necessary results for a detailed analysis of the sediment transport in a specific irrigation canal or network some distinct and clear details and steps are required. Based on a thorough analysis of a specific sediment problem and its alternatives the following data are required before the start of the simulation, namely the physical geometry of the canal and network, the characteristics of the incoming sediment and the details of the required calculation steps, namely for the time and length step. The details of the data are given in Table 7.11.

Based on the above described data the model will calculate the hydraulic variables by the predictor-corrector method per time step and length step; after each time step the sediment transport is calculated. The hydraulic calculation will give the roughness, water depth and velocity for each time step and at each point. Next the model calculates the actual sediment characteristics based upon one of the three sediment predictors (Ackers-White, Brownlie or Engelund-Hansen). The adaptation time T_A and length L_A of the incoming sediment load (c_0) produce the equilibrium sediment transport, which is given by the sediment concentration from Gallapatti's model that is based on the 2-D convection-diffusion equation,

Table 7.11. Data required by SETRIC before the start of any simulation of the hydraulic and sediment calculations.

Physical geometry of the canal and network includes:
- layout of the canal network with main and secondary canals (laterals), number of canal sections, location of control and other structures, offtakes, bifurcations and confluences;
- chainage and length of the canal sections;
- for each section the bottom width, roughness, bottom slope, side slope, starting bottom level and drops;
- base flow and discharge per lateral and the roughness (Chézy, Manning, Strickler);
- maintenance activities, characteristics for the weed factors and other constants;
- number, location and type of structures: overflow, flume, siphon-culvert and undershot;
- dimensions of the structures;
- total number cannot be more than the number of sections;
- water depth upstream of the control structure at the downstream boundary.

Characteristics of the water flow
- base flow (discharge) at the upstream boundary;
- operation mode, either continuous or intermittent;
- type of canal maintenance; none, good or ideal maintenance;
- lateral flow and location of the laterals from the head of the parent canal;
- flow options for the laterals, either continuous or in time series;
- option for simulation of sediment in the lateral; laterals will get the same sediment characteristics as the parent canal;
- only 1 lateral can start from 1 node;
- geometrical dimensions to be included in the same way as for the parent canal.

Characteristics of the incoming sediment include:
- incoming sediment is non-colloidal;
- sediment particles are characterized by 1 size (d_{50});
- size and gradation of the sediment constant along the whole canal length, either continuous or in time series;
- total concentration (bed load and suspended) in ppm by weight; either continuous or in time series;
- type of sediment predictor: Ackers-White, Brownlie or Engelund-Hansen;
- no sediment deposition in the structures, the sediment inflow is equal to the sediment outflow;
- only previously deposited sediment can be eroded, the original canal bed is different from the inflowing sediment; cannot be eroded.

Details of the required calculation steps should be known before the simulation:
- time step Δt has to be known before the simulation starts, Δt is a function of the adaptation time T_A and is also related to the Courant number; length step Δx depends on the required accuracy for the numerical solution and should be much smaller than the adaptation length L_A and should be smaller than 0.67 times the smallest length of a reach with structures at both nodes.

which gives the actual sediment concentration at any point under non-equilibrium conditions. The mass balance equation for the total sediment transport is solved by Lax's modified method, assuming steady conditions for the sediment concentration.

The actual (non-equilibrium) sediment concentration depends on c_0, L_A, T_A, Δx and c_e. The value of c_0 depends on the inflowing sediment concentration at the headworks. At the boundaries between canal reaches the sediment load passing the downstream boundary is c_0 for the next reach. The depth-averaged equilibrium concentration c_e is based upon one of the three sediment transport predictors. The predictors give the sediment transport per unit width, and the bottom width B and the correction factor α are required to determine the total sediment transport, which will form the basis for the equilibrium concentration. The model simulates the total sediment load, namely the bed load and suspended load together. If a structure has a raised crest then it might trap a part of the bed load upstream of the structure. Only the whole suspended load part will move to the downstream canal reach. Internal conditions such as bifurcations or confluences are needed for continuity of water flow and sediment transport to find changes either in bottom width or bottom level.

During the first time step of the simulation process the water flow is uniform or develops a gradually varied flow. For these conditions the sediment characteristics will be determined, especially the equilibrium sediment concentration. When the actual sediment concentration differs from the equilibrium concentration, the latter will try to adapt to the equilibrium concentration by deposition of the sediment in the water or by erosion of sediment from the bottom. Due to the sedimentation and erosion the hydraulic characteristics will change and hence the water flow variables for the next time step will change. A new gradually varied flow will be established, resulting in a quasi-steady flow. In this step the actual sediment concentration will again adapt to the equilibrium concentration, resulting in sedimentation or erosion, which will lead to changes in the geometrical characteristics of the canal, which will be used for the sediment calculations in the next time step. This process will continue either until the actual sediment concentration is in line with equilibrium concentration or will continue until the end of the simulation period.

A reflection of the previously presented examples and simulation results shows that the SETRIC model is a useful tool to develop a better understanding of the behaviour of sediment transport in irrigation canals and to decide on improved water delivery plans in view of the operation and maintenance concepts under different flow conditions and sediment characteristics. It can be applied to evaluate designs of an irrigation network and to analyse the alternatives, but it can also be used as a decision support tool in the operation and maintenance of a system and/or to determine the efficiency of sediment removal facilities in an irrigation system. In spite of that, the mathematical model still has to be calibrated and validated with field data before the model can be fully used as a simulation tool. The performance of the model has to be confirmed by field measurements that should prove whether the physical processes are well represented by the model or whether there are some deficiencies as a result of the assumptions used to describe those processes. Monitoring of the sediment deposition in an irrigation network is needed to evaluate the model and to investigate the response of the bottom level in time and space to determine water flows and sediment characteristics. Influences such as the type and operation of flow control structures, geometrical characteristics of the canals, water flow and the incoming sediment characteristics on the deposition, which the mathematical model predicts, will contribute to a better understanding of the sediment transport processes under the flow conditions prevailing in a specific irrigation canal or network.

The model forms a sound basis to evaluate design alternatives for minimal sedimentation and erosion, to maintain a fair, adequate and reliable water supply to the farmers, and to decide on the applicability of these concepts for the simulation of the sediment transport processes under the particular conditions of water flow and sediment inputs.

References

Abbot, M.B. & J.A. Cunge, 1982. *Engineering applications of compu-tational hydraulics, Volume I.* Pitman Books Limited, Massachusetts, USA.

Ackers, P. & W.R. White, 1973. Sediment transport: new approach and analysis. *Journal of Hydraulic Division, ASCE.* Vol. 99, No. 11. New York, USA.

Ackers, P., 1992. Gerard Lacey memorial lecture: Canal and river regime in theory and practice: 1929–92. *Proc. Institution Civil Engineers. Water, Maritime & Energy.* Technical note No. 619. United Kingdom.

Ankum, P., 1995. *Lectures notes on flow control in irrigation and drainage.* UNESCO-IHE. Delft, the Netherlands.

Ankum, P., 2004. *Lectures notes on Flow Control in Irrigation Systems.* UNESCO-IHE. Delft, the Netherlands.

Armanini, A. & G. Di Silvio, 1988. A one dimensional model for the transport of a sediment mixture in non-equilibrium condition. *Journal of Hydraulic Research, IAHR.* Vol. 26, No. 3. The Netherlands.

Asano, T. et al., 1985. Characteristics of variation of Manning's rough-ness coefficient in a compound cross section. *21st IAHR congress.* Melbourne, Australia.

ASCE, 1966. Initiation of Motion. Task committee on preparation of sedimentation manual. *Journal of Hydraulic Division, ASCE.* New York, USA.

Bagnold, R., 1966. An approach to the sediment transport problem from general physics. *Geological Survey Prof. Paper 422.I.* Washington, USA.

Bakhmeteff, B.A., 1932. *Hydraulics of open channels.* McGraw-Hill Book Company, New York, USA.

Bakker, B. et al., 1989. Regime theories updated or outdated. *Delft Hydraulics.* Publication no. 416. Delft, the Netherlands.

Baume, J.-P., P.-O. Malaterre, G. Belaude & B.L. Guennec, 2005. SIC: A 1D hydrodynamic model for river and irrigation canal modeling and regulation. User manual of SIC model. CEMAGREF, Montpellier, France.

Bettes, R., W.R. White & C.E. Reeve, 1988. Width of regime channels. International conference on river regime. Hydraulic research.
Wallingford, Great Britain.

Bishop, A. et al., 1965. Total bed material transport. *Journal of Hydraulic
Division, ASCE.* Vol. 91, No. 2. New York, USA.

Blench, T., 1957. Regime behaviour of canals and rivers. Butterworth's
Scientific Publications. London, Great Britain.

Blench, T., 1970. Regime theory design of canals with sand beds.
Journal of Irrigation and Drainage Division, ASCE. Vol. 96, No. 2.
New York, USA.

Bogardi, J., 1974. *Sediment transport in alluvial streams.* Akademiai
Kiado. Budapest, Hungary.

Bos, M.G. (Ed.), 1989. *Discharge measurement structures* (Third edition). International Institute for Land Reclamation and Improvement.
Wageningen, The Netherlands.

Brabben, T., 1990. Workshop on Sediment Measurement and Control,
and Design of Irrigation Canals. Hydraulic Research. Wallingford,
Great Britain.

Brebner, A. & K.C. Wilson, 1967. Determination of the Regime Equation from Relationship for Pressurised Flow by Use of the Principle
of Minimum Energy Degradation. *Proceedings Institutions of Civil
Engineers.* Vol. 36, Issue 1, pp. 47–62. London, Great Britain.

Breuser, H.N., 1993. *Lecture notes on sediment transport.* UNESCO-IHE.
Delft, the Netherlands.

Brownlie, W., 1981a. Compilation of alluvial channel data: laboratory
and field. *California Institute of Technology.* Report No. KH-R-43B.
California, USA.

Brownlie, W., 1981b. Prediction of flow depth and sediment discharge
in open channels. California Institute Technology. W.M. Keck Lab.
Rep. No. KH-R-54. Laboratory of Hydraulic and Water Resources,
California, USA.

Brownlie, W., 1983. Flow depth in sand bed channels. *Journal of
Hydraulic Division, ASCE.* Vol. 109, No. 7. New York, USA.

Bruk, S., 1986. Sediment Transport and Control in Irrigation System.
*International Conference on water Resources Needs and Planning in
Drought Prone Area.* Khartoum, Sudan.

Bureau of Reclamation, 1987. *Design of small dams.* U.S. Department of
the Interior Bureau of Reclamation. US Government Printing Office,
Washington DC, USA.

Bureau of Reclamation, 2001. *Water Measurement Manual: A Guide
to Effective Water Measurement Practices for Better Water Management.* U.S. Department of the Interior Bureau of Reclamation, U.S.
Government Printing Office, Washington DC, USA.

Chang, H., 1967. *Hydraulics of rivers and deltas.* PhD dissertation.
Colorado State University. Colorado, USA.

Chang, H., 1980. Stable Alluvial Canal Design. *Journal of Hydraulic Division, ASCE*, Vol. 106, No. 5, pp. 873–891. New York, USA

Chang, H., 1985. Design of stable alluvial canals in a system. *Journal of Hydraulic Division, ASCE.* New York, USA.

Chang, H., 1988. *Fluvial processes in river engineering.* John Wiley & Sons. New York, USA.

Chitale, S.V., 1994. Lacey divergence equation for alluvial canal design. *Journal of Hydraulic Division, ASCE.* Vol. 120, No. 1. New York, USA.

Chow, Ven Te, 1983. *Open channel hydraulics.* McGraw Hill International Book Company. Tokyo, Japan.

Chuang Yang, et al., 1989. Semi-coupled Simulation of Unsteady Non uniform Sediment Transport in Alluvial Canal. *23rd IAHR congress.* Ottawa, Canada.

Clemmens, A.J., F.M. Holly & W. Schuurmans, 1993. Description and evaluation of Duflow. *Journal of Hydraulic Division, ASCE.* Vol. 119, No. 4. New York, USA.

Clemmens, A.J., T. Wahl, M. Bos et al., 2001. *Water Measurement with Flumes and Weirs.* ILRI Publication No 58. International Institute for Land Reclamation and Improvement. Wageningen, The Netherlands.

Colby, B., 1964. Practical computations of bed material discharge. *Journal of Hydraulic Division, ASCE.* Vol. 90, No. 2. New York, USA.

Cunge, J.A., F.M. Holly & A. Verwey, 1980. *Practical Aspects of Computational River Hydraulics.* Pitman Advanced Publishing Program. London, Great Britain.

Dahmen, E.R., 1994. *Lecture notes on canal design.* UNESCO-IHE. Delft, the Netherlands.

Dahmen, E.R., October 1999. *Irrigation and Drainage Systems Part 1*, Lecture Notes of Hydraulic Engineering, UNESCO-IHE. Delft, the Netherlands.

Davies, T.R.H., 1982. Lower flow regime bed forms: rational classification. *Journal of Hydraulic Division, ASCE.* Vol. 108, pp. 343–360. New York, USA.

Depeweg, H., 1993. *Lecture notes on applied hydraulics: gradually varied flow.* UNESCO-IHE. Delft, the Netherlands.

Depeweg, H., 2000. *Lecture notes on structures in irrigation networks.* UNESCO-IHE. Delft, the Netherlands.

Depeweg, H. & V. Néstor Méndez, 2002. Sediment Transport Applications in Irrigation Canals. *Irrigation and Drainage Journal* 51: pp. 167–179. John Wiley & Sons Ltd. New York, USA.

Depeweg, H. & K.P. Paudel, 2003. Sediment Transport Problems in Nepal Evaluated By the SETRIC Model. *Irrigation and Drainage Journal* 52: pp. 247–260. John Wiley & Sons Ltd. New York, USA.

Depeweg, H. & E. Rocabado Urquieta, 2004. GIS Tools and the Design of Irrigation Canals. *Irrigation and Drainage* 53: pp. 301–314. John Wiley & Sons Ltd. New York, USA.

DHL (Delft Hydraulics Laboratory) & Ministry of Transport, Public Works and Water Management, 1994. *SOBEK: technical reference guide.* Delft Hydraulics. Delft, the Netherlands.

Einstein, H.A. & N.L. Barbarossa, 1952. River channel roughness. *ASCE Transactions*, Volume 117, Paper 2528, pp. 1121–1146. New York, USA.

Einstein, H.A., 1942. Formulas for the transportation of bed-load. *ASCE Transactions* Vol. 107. pp 133–169. New York, USA.

Einstein, H.A., 1950. The bed load function for sediment transportation in open channel flow. *Bulletin No. 1026*, U.S. Dep. of Agriculture, Washington, USA.

Engelund, F. & E. Hansen, 1967. *A monograph on sediment transport in alluvial streams.* Teknisk Forlag, Copenhagen, Denmark.

Engelund, F., 1966. Hydraulic resistance of alluvial stream. *Journal of Hydraulic Division, ASCE.* Vol. 92, No. 2. New York, USA.

FAO, 1981. Arid zone hydrology. *Paper No. 37, FAO.* Rome, Italy.

FAO, 1992. Wastewater treatment and use in agriculture – Irrigation and drainage paper 47, FAO – Food and Agriculture Organization of the UN, Rome.

FAO, 1998. Crop evapotranspiration – Guidelines for computing crop water requirements Irrigation and drainage paper No. 56, FAO – Food and Agriculture Organization of the UN, Rome.

FAO, 2003. The irrigation challenge: Increasing irrigation contribution to food security through higher water productivity from canal irrigation systems. *Issues Paper No. 4, FAO*, Rome, Italy.

Galappatti, R., 1983. A Depth Integrated Model for Suspended Transport. *Report no 83-7. Delft University of Technology.* Delft, the Netherlands.

Ghimire, P.K., 2003. Non-equilibrium Sediment Transport in Irrigation Canals. *M.Sc. Thesis UNESCO-IHE.* Delft, The Netherlands.

Henderson, F.M., 1966. *Open Channel Flow.* MacMillan Publishing Co. Inc. New York, USA.

HR Wallingford, 1990. Sediment Transport, The Ackers and White Theory Revised. No. SR237, HR Wallingford, England.

HR Wallingford, 1992. *DORC: user manual.* HR Wallingford. Wallingford, Great Britain.

IHE, 1998. *Duflowmanual.* UNESCO-IHE. Delft, the Netherlands.

Ida, Y., 1960. Steady flow in wide channel and the effect of shape of its cross section. (in Japanese) Trans. Japan Society of Civil Engineers. *Journal of JSCE, Division B: Hydraulic, Coastal and Environmental Engineering*, Tokyo, Japan.

Ikeda, S., 1982a. Lateral bed load transport on side slopes. *Journal of Hydraulic Division*, Vol. 108, No. HY11. New York, USA.

Ikeda, S., 1982b. Incipient motion of sand particles on side slopes. *Journal of Hydraulic Division*, Vol. 108, No. HY1, ASCE. New York, USA.

Jansen, P., 1994. *Principles of river engineering.* Delftse Uitgevers Maatsschappij. Delft, The Netherlands.

Kennedy, R.G., 1895. The prevention of silting in irrigation canals. *Minutes of the Proceedings.* Vol. 119, Issue 1895, pp. 281–290. London, Great Britain.

Kerssens, P.J.M., A. Prins & L.C.v. Rijn, 1979. Model for Suspended Sediment Transport. *Journal of Hydraulic Division, ASCE,* Vol. 105 (HY5). New York, USA.

Kinori, B. Z., 1970. *Manual of Surface Drainage Engineering.* Volume 1 Developments in Civil Engineering. Elsevier Science Ltd.

Klaassen, G.J., 1995. Lane's Balance Revisited. *Proc. 6th International Symposium on River Sedimentation.* New Delhi, India.

Kouwen, N., 1969. Flow retardance in vegetated channels. *Journal of Irrigation and Drainage Division, ASCE.* Vol. 95, No. 2. New York, USA.

Kouwen, N., 1988. Field estimation of biomechanical properties of grass. *Journal of Hydraulic Research, ASCE.* Vol. 26, No. 5. pp. 559–568. New York, USA.

Kouwen, N., 1992. Modern approach to design of grassed channels. *Journal of Hydraulic Division, ASCE.* Vol. 118, No. 5. New York, USA.

Krishnamurthy, M. & B, Christensen, 1972. Equivalent roughness for shallow channels. *Journal of Hydraulic Division, ASCE.* Vol. 118, No. 5. New York, USA.

Krüger, F., 1988. *Flow laws in open channels.* Doctoral thesis at the Dresden University of Technology. Dresden, Germany.

Lacey, G., 1930. Stable channels in alluvium. *Proceedings Institutions of Civil Engineers.* Vol. 229, pp. 259–292. London, Great Britain.

Lacey, G., 1958. Flow in alluvial channels with sandy mobile beds. *Proceedings of the Institution of Civil Engineers,* 9: 145–164. London, United Kingdom.

Lane, E.W. & A.A. Kalinske, 1941. Engineering calculations of suspended sediment. *Transactions American Geophysical Union 22,* Washington, USA.

Lane, E.W., 1953. Progress report on studies on the design of stable channels. Bureau of Reclamation. *Proceedings ASCE 79,* New York, USA.

Lane, E.W., 1955. Design of stable alluvial channels. *Transactions, ASCE,* Vol. 120. pp. 1234–1260. New York, USA.

Lawrence, E.F. & Miguel Marino, 1987. Canal design: optimal cross section. *Journal of Hydraulic Division, ASCE.* Vol. 113, No. 3. New York, USA.

Lawrence, P., 1990. Canal Design, Friction and Transport Predictor. Workshop on Sediment Measurement and Control, and Design of Irrigation Canals. Hydraulic Research. Wallingford, Great Britain.

Lawrence, P., 1993. Deposition of Fine Sediments in Irrigation Canals. *15th ICID congress.* The Hague, the Netherlands.

Leliavsky, S., 1983 *Irrigation Engineering: canals and barrages*, Chapman and Hall. London, Great Britain.

Lier, H.N. van, L.S. Pereira & F.R. Steiner, 1999. *CIGR handbook of agricultural engineering, Volume 1*: Land and water engineering. American Society of Agricultural Engineers. St. Joseph, Michigan, USA.

Lindley, E.S., 1919. Regime Channels. *Proceeding Punjab Engineering Congress.* Vol. 7, pp. 63–74. Lahore, Pakistan.

Liu, H.K., 1957. Mechanics of ripple formation. *Journal of Hydraulic Division.* Proceedings ASCE. New York, USA.

Lyn, D.A., 1987. Unsteady Sediment Transport Modelling. *Journal of Hydraulic Division, ASCE.* Vol. 113. New York, USA.

Mahmood, K., 1973a. Sediment Equilibrium Consideration in The Design of Irrigation Canal Network. *Proc. IAHR. International Symposium on River Mechanics.* Bangkok, Thailand.

Mahmood, K., 1973b. Sediment routing in irrigation canal systems. *ASCE,National Water Resources Engineering Meeting.* Washington DC, USA.

Mahmood, K. & V. Yevjevitch, 1975. *Unsteady flow in open channels.* Volume 1, Water Resources Publications, Fort Collins, Colorado.

Méndez, N.V., 1995. Suspended sediment transport in irrigation canals. *M.Sc thesis. UNESCO-IHE.* Delft, The Netherlands.

Méndez, N.V., 1998. *Sediment Transport in Irrigation Canals.* A.A. Balkema. Rotterdam, The Netherlands.

Merkley, G.P., 1997. CanalMan: A Hydraulic Simulation Model for Unsteady Flow in Branching Canal Networks, *User's Guide.* Dept. of Biological and Irrigation Engineering, Utah State University, Utah, USA.

Motayed, A. & M. Krishnamurthy, 1980. Composite roughness of natural channels. *Journal of Hydraulic Division, ASCE.* Vol. 106, No. HY6. New York, USA.

Munir, S., 2011. *Role of Sediment Transport in Operation and Maintenance of Supply Based and Demand Based Irrigation Canals.* Application to the Macha Maira Branch Canals. CRC Press/Balkema. Leiden, The Netherlands.

Nagakawa, H. et al., 1989. Convolution-integral modelling of non-equilibrium sediment transport. *Sediment transport modelling. Ed. by S. S. Y. Wang. ASCE.* New York, USA.

Naimed Ullah, M., 1990. Regime and Sediment Transport Concepts Compared as Design Approaches. *Workshop on Sediment Measurement and Control, and Design of Irrigation Canals.* Hydraulic Research. Wallingford, Great Britain.

Nitschke, E., 1983. The influence of overgrowing with herbs on hydraulics parameters of agricultural outfalls and ditches. *Proc. 20th Congress of IAHR.* Subject D, pp. 173–180. Moscow, USSR.

Ogink, H.J.M., 1985. The effective viscosity coefficient in 2-D depth-averaged flow models. *Proc. 21st IAHR congress.* Melbourne, Australia.

Paintal, A.S., 1971. Concept of critical shear stress in loose boundary open channel. *IAHR. No. 9.* Delft, the Netherlands.

Paudel, K.P., 2002. Evaluation of the sediment transport model SETRIC for irrigation canal. *MSc thesis. UNESCO-IHE.* Delft, the Netherlands

Paudel, K.P., 2010. *Role of Sediment in the Design and Management of Irrigation Canals*, Sunsari Morang Irrigation Scheme, Nepal. CRC Press/Balkema. Leiden, The Netherlands.

Paudel, K., B. Schultz & H. Depeweg, 2013. Modelling of an irrigation scheme with sediment-laden water for improved water management: a case study of Sunsari Morang Irrigation Scheme, Nepal. *ICID journal.* New Delhi, India.

Querner, E., 1993. *Aquatic weed control within an integrated water management framework.* Doctoral thesis, Agricultural University of Wageningen. Wageningen, the Netherlands.

Rajaratnam, N. & K. Subramanya, 1967. Flow equation for the sluice gate. *Journal of Irrigation and Drainage Division*, ASCE, 93(3): 167–186.

Ranga Raju, K.G., K. Dhandapani & D. Kondup, 1977. Effect of Sediment Load on Stable Sand Canal Dimensions. *Journal of Waterway, Port, Coastal and Ocean Division*, ASCE, 103(WW2): 241–249.

Ranga Raju, K.G., 1981. *Flow through open channels.* Tata McGraw-Hill. New Delhi, India.

Raudkivi, A.J. & H. Witte, 1990. Development of bed features. *Journal of Hydraulic Division, ASCE.* Vol. 116, No. 9. New York, USA.

Raudkivi, A.J., 1990. *Loose Boundary Hydraulics.* 3rd Edition. Pergamon Press. Great Britain.

Rhodes, D.G., 1995. Newton-Raphson Solution for Gradually Varied Flow. *Journal of Hydraulic Research*, Vol. 33, 1995, No. 2, pp. 213–218. New York, USA.

Ribberink, J.S., 1986. Introduction to a depth integrated model for suspended sediment transport (Galappatti, 1983). *Report No. 6–86. Delft University of Technology.* Delft, the Netherlands.

Rijn, L.C. van, 1982. Equivalent roughness of alluvial bed. *Journal of Hydraulic Division, ASCE.* New York, USA.

Rijn, L.C. van, 1984a. Sediment Transport Part I: Bed load Transport. *Journal of Hydraulic Division, ASCE.* Vol. 110, No. 10. New York, USA.

Rijn, L.C. van, 1984b. Sediment Transport Part II: Suspended Load Transport. *Journal of Hydraulic Division, ASCE.* Vol. 110, No. 11. New York, USA.

Rijn, L.C. van, 1984c. Sediment transport part III: Bed form and alluvial roughness. *Journal of Hydraulic Division, ASCE.* Vol. 110, No. 12. New York, USA.

Rijn, L.C. van, 1993. *Lecture notes on principles of sediments transport in rivers, estuaries, coastal seas and oceans.* UNESCO-IHE. Delft, the Netherlands.

Schuurmans, W. 1991. *A model to study the hydraulic performance of controlled irrigation canals.* Doctoral thesis. Delft University of Technology. Delft, the Netherlands.

Shen, H.W., 1976. *Stochastic approaches to water resources* (Vol. I and II). Colorado State University. Fort Collins, Colorado, USA.

Sherpa, K., 2005. Use of the Sediment Transport Model Setric in an Irrigation Canal. *MSc Thesis.* UNESCO-IHE, the Netherlands.

Simons, D. & F. Senturk, 1992. *Sediment Transport Technology.* Water Resources Publications. Colorado, USA.

Simons, D.B. & E.V. Richardsons, 1961. Forms of bed roughness in alluvial channels. *Journal of Hydraulic Division, ASCE.* No. 3. New York, USA.

Simons, D.B. & M.L. Albertson, 1963. Uniform Water Conveyance Channels in Alluvial Material. *Transactions, ASCE*, Vol. 128, No. 1. pp. 65–167. New York, USA.

Stevens, M.A. & Nordin, C.F., 1987. Critique of the Regime Theory for Alluvial Channels. *Journal of Hydraulic Engineering, ASCE*, 113(11): 1359–1380.

Timilsina, P., 2005. One-Dimensional Convection Diffusion Model for Non-Equilibrium Sediment Transport in Irrigation Canals. *MSc Thesis.* UNESCO-IHE, the Netherlands.

Toffaletti, F.B., 1969. Definitive computations of sand discharge in rivers. *J. of the Hydraulic Div. ASCE*, Vol. 95, No. HY1, January, pp. 225–246.

UNESCO-IHE, 1998. *Duflow manual.* UNESCO-IHE. Delft, the Netherlands.

Utah State University, 2006. RootCanal User's Guide. Dept. of Biological and Irrigation Engineering, Utah State University. Utah, USA.

Vanoni, V. & N.H. Brooks, 1957. Laboratory Studies of the Roughness and Suspended Load of Alluvial Streams. *No. E-68, California Institute of Technology.* Pasadena, California, USA.

Vanoni, V. (Ed.), 1975. Sedimentation Engineering. *Manuals and reports on engineering practice No. 54. ASCE.* New York, USA.

Vlugter, H., 1962. Sediment transportation by running water and design of stable alluvial channels. *De Ingenieur.* The Hague, The Netherlands.

Vos, H.C.P. de, 1925. Transport of solid materials by flowing water. (in Dutch). *Waterstaats Ingenieur*, Banjuwangi, Indonesia

Vreugdenhil, C.B., 1989. *Computational Hydraulics*. Springer-Verlag. Berlin, Germany.

Vreugdenhil, C.B. & J.H.A. Wijbenga, 1982. Computation of Flow Patterns in Rivers. *Journal of Hydraulic Division, ASCE*, 108(11): 1296–1310.

Vries, M. de, 1965. Consideration about non-steady bed load transport in open channel. Delft Hydraulics Laboratory. Publications No. 36. Delft, The Netherlands.

Vries, M. de, 1975. A morphological time scale for rivers. *IAHR*. Sao Paulo, Brazil.

Vries, M. de, 1985. A Sensitivity Analysis Applied to Morphological Computations. *Report No. 85-2. Delft University of Technology.* Delft, the Netherlands.

Vries, M. de, 1987. *Lecture notes on Morphological Computations.* Delft University of Technology. Delft, the Netherlands.

Walker, W.R., 1993. *SIRMOD, Surface irrigation simulation software.* Utah State University. Logan, Utah, USA.

Wang, Z.B. & Ribberink, J.S., 1986. The validity of a depth integrated model for suspended sediment transport. *Journal of Hydraulic Research, IAHR.* Vol. 24, No. 1. Delft, the Netherlands.

White, W.R. et al., 1979. A new general method for predicting the frictional characteristics of alluvial streams. *Hydraulic Research Limited*, Wallingford, Great Britain.

White, W.R., E. Paris & R. Bettes, 1979. General Method for Predicting the Functional Characteristics of Alluvial Streams. *Hydraulics Research Station Report* No IT 187, Wallingford, England.

White, W.R., R. Bettes & E. Paris, 1980. The frictional characteristics of alluvial stream; A new approach. *Proceedings of the Institution of Civil Engineers*, 69(2): 737–750.

White, W.R., R. Bettes & E. Paris, 1981. Tables for the design of stable alluvial channels. *No. IT 208, Hydraulics Research Station*, Wallingford, Great Britain.

Woo, H. & K. Yu, 2001. Reassessment of Selected Sediment Discharge formulas. *Proc. XXIX IAHR Congress.* Beijing, China. 224–230.

Worapansopak, J., 1992. Control of Sediment in Irrigation Schemes. *MSc thesis. UNESCO-IHE.* Delft, the Netherlands.

World Bank, 1986. Design and operating guidelines for structural irrigation networks. *Fourth draft. Irrigation II Division. South Asia Project Department. World Bank.* Washington DC, USA.

World Bank, 1995. The World Bank and Irrigation. *World Bank Operation Evaluation Department. The World Bank.* Washington DC, USA.

Yalin, M.S., 1977. *Mechanics of Sediment Transport.* Pergamon Press. Oxford, Great Britain.

Yalin, M.S., 1985. On the determination of ripple geometry. *Journal of Hydraulic Division, ASCE.* New York, USA.

Yang, C.T. & A. Molinas, 1982. Sediment transport and Unit stream power. *Journal of Hydraulic Division 40.* Amsterdam, the Netherlands.

Yang, C.T. & S. Wan, 1991. Comparisons of Selected Bed-Material Load Formulas. *Journal of Hydraulic Engineering, ASCE.* Vol.117. No. 8, pp. 973–989. New York, USA.

Yang, C.T., 1973. Incipient motion and sediment transport. *Journal of Hydraulic Division, ASCE.* Vol. 99. No. 110. New York, USA.

Yang, C.T., 1979. Unit stream power equation for total load. *Journal of Hydrology Division, ASCE.* Vol. 99. No. 110. New York, USA.

Yang, C.T., C. Yong & M. Woldenberg, 1981. Hydraulic Geometry and Minimum Rate of Energy Dissipation. Water Resources Research, 17(4): 1014–1018. USA.

Yang, S.Q. & S.Y. Lim, 1997. Mechanism of energy transportation and turbulent flow in a 3D channel. *Journal of Hydraulic Engineering, ASCE,* Vol. 123. No. 8, pp. 684–692. New York, USA.

Yang, S.Q., J.X. Yu & Y.Z. Wang, 2004. Estimation of diffusion coefficients, lateral shear stress and velocity in open channels with complex geometry. *Water Resources Research, 40*(W05202).

Yen, B.C., 2002. Open Channel Flow Resistance. *Journal of Hydraulic Engineering, ASCE,* Vol. 128. No. 1, pp. 20–39. New York, USA.

Symbols

SI-UNITS

Basic units

Quantity	Unit	SI Symbol
length	meter	m
mass	kilogram	kg
time	second	s
plane angle	radian	rad

Derived units

Quantity	Unit	SI Symbol
acceleration	meter per second squared	m/s^2
area	square meter	m^2
density	kilogram per cubic meter	kg/m^3
discharge	cubic meter per second	m^3/s
energy	Joule	$J = N \cdot m$
force	Newton	$N = kg \cdot m/s^2$
pressure	Pascal	$Pa = N/m^2$
power	Watt	$W = Nm/s$
velocity	meter per second	m/s
viscosity, dynamic	Pascal-second	$Pa \cdot s$
viscosity, kinematic	square meter per second	m^2/s
volume	cubic meter	m^3

LIST OF SYMBOLS

Symbol	Quantity	Unit	Dimension
A	Area, cross section area	(m^2)	L^2
a	Height of half roughness $k/2$	(m)	L
B	Channel bottom width	(m)	L
B_o	Channel width; bed width	(m)	L
B_s	Water surface width	(m)	L
B_s	Sediment transport width	(m)	L
B_s	Water surface width	(m)	L
C	de Chézy resistance coefficient	(m$^{1/2}$/s)	L$^{1/2}$T^{-1}
C'	de Chézy coefficient due to skin resistance	(m$^{1/2}$/s)	L$^{1/2}$T^{-1}
C''	de Chézy coefficient due to form resistance	(m$^{1/2}$/s)	L$^{1/2}$T^{-1}
c	Sediment concentration	(%, ppm by mass)	–
C	Sediment concentration	(%, ppm by mass)	–
c_0	Concentration at length $x = 0$	(%, ppm by mass)	–
C_0	Total sediment concentration at $x = 0$	(%, ppm by mass)	–
c_a	Reference concentration at level 'a' from the bottom	(%, ppm by mass)	–
c_b	Bed load concentration	(%, ppm by mass)	–
C_c	Contraction coefficient	(–)	–
C_d	Discharge coefficient	(–)	-
C_D	Drag coefficient	(–)	–
c_e	Equilibrium concentration	(%, ppm by mass)	–
C_e	Total equilibrium sediment concentration	(%, ppm by mass)	–
C_e	Effective Chézy coefficient	(m$^{1/2}$/s)	L$^{1/2}$T^{-1}
C_e'	Modified effective Chézy coefficient	(m$^{1/2}$/s)	L$^{1/2}$T^{-1}
C_f	Friction coefficient, $(u_*/v)^2$	(–)	–
c_t or C_t	Total sediment concentration	(%, ppm by mass)	–
C_v	Velocity coefficient	(–)	–
d	Particle diameter	(m)	L
D	Diameter	(m)	L
$d*$	Dimensionless particle diameter	(–)	–
D_*	Dimensionless particle parameter	(–)	–
d_{50}	Median particle diameter	(m)	L
d_i	Diameter of i-percent finer than d_i	(m)	L
d_i	Representative particle diameter	(m)	L
d_r	Discrepancy ratio	(–)	–
E	Total energy	(Nm/N)	L
e	Exponential number; e $= 2.71828$	(–)	–
E	Total energy in relation to a horizontal datum	(J/N)	
E_s	Specific energy in relation to lowest point of a cross-section	(Nm/N)	L
f	Darcy-Weisbach resistance coefficient	(–)	–
f	Friction factor	(–)	–
f	Lacey's silt factor	(–)	–
F	Shape factor	(–)	–
$f(\ldots)$	Function of	(–)	–
F_g	Grain Froude number	(–)	–
F_g	Weed factor	(–)	–
F_{gcr}	Critical grain Froude number	(–)	–
F_{gr}	Dimensionless mobility parameter	(–)	–

(Continued)

Symbol	Quantity	Unit	Dimension
Fr	Froude number	(−)	−
g	Acceleration due to gravity	(m/s^2)	LT^{-2}
G_{gr}	Dimensionless mobility parameter	(−)	−
h	Water depth	(m)	L
K	Error factor	(−)	−
k	Roughness height	(m)	L
k_s	Roughness height of Nikuradse	(m)	L
k'_s	Grain roughness height	(m)	L
k''_s	Bed form roughness height	(m)	L
k_{se}	Effective equivalent roughness	(m)	L
k_s	Strickler coefficient	(m$^{1/3}$/s)	L$^{1/3}$T^{-1}
$k_{m,d}$	modified k_m for a canal depth d (Strickler)	(m$^{1/3}$/s)	L$^{1/3}$T^{-1}
$k_{m,1m}$	reference k_m for a canal depth of 1.0 m (Strickler)	(m$^{1/3}$/s)	L$^{1/3}$T^{-1}
l, L	Length	(m)	L
L_A	Adaptation length	(m)	L
m	Cotangent of side slope	(−)	
m	Side slope (1:m)	(−)	
M	Momentum function	(−)	
n	Manning resistance coefficient	(s/m$^{1/6}$)	L$^{1/6}$
n_e	Equivalent Manning coefficient	(s/m$^{1/6}$)	L$^{1/6}$
P	Wetted perimeter	(m)	L
p	Porosity	(−)	−
P	Power	(W)	
Q	Discharge; water discharge, flow discharge	(m^3/s)	L^{3}T^{-1}
q	Discharge per unit width of flow	(m^3/s/m)	L^{2}T^{-1}
q_*	Dimensionless unit discharge	(−)	−
q_b	Bed load transport rate per unit width	(m^3/s/m)	L^{2}T^{-1}
q_e	Equilibrium sediment discharge	(m^3/s/m)	L^{2}T^{-1}
q_s	Volumetric rate of sediment transport per unit width of flow	(m^3/s/m)	L^{2}T^{-1}
q_s	Sediment discharge per unit width	(m^3/s/m)	L^{2}T^{-1}
Q_s	Total sediment discharge	(m^3/s)	L^{3}T^{-1}
Q_{so}	Sediment discharge at upstream point	(m^3/s)	L^{3}T^{-1}
q_{st}	Flow discharge in a stream tube	(m^3/s)	L^{3}T^{-1}
q_{sus}	Suspended load transport per unit width	(m^3/s/m)	L^{2}T^{-1}
r	Radius	(m)	L
R	Hydraulic radius	(m)	L
Re	Reynolds number	(−)	−
Re$_*$	Particle Reynolds number	(−)	−
R_g	Grain Reynolds number	(−)	−
R_v	Velocity ratio	(−)	−
S	Slope	(−)	−
s	Relative density	(−)	−
s.f.	Shape factor	(−)	−
S_f	Friction slope	(−)	−
S_o	Bed slope	(−)	−
t	Time coordinate	(s)	T
T	Excess bed shear stress parameter	(−)	−
T_A	Adaptation time	(s)	T
T/Q	Relative transport capacity	(−)	−

(Continued)

Symbol	Quantity	Unit	Dimension
u_*	Shear velocity	(m/s)	LT^{-1}
u_{*cr}	Critical shear velocity	(m/s)	LT^{-1}
u, v, w	Component of velocity in the X, Y and Z direction	(m/s)	LT^{-1}
v	Local velocity in x direction	(m/s)	LT^{-1}
V	Volume	(m^3)	L^3
V	Mean velocity of flow	(m/s)	LT^{-1}
v_*	Shear velocity	(m/s)	LT^{-1}
v_{*C}	Critical value of v_* associated with incipient motion of bed particles	(m/s)	LT^{-1}
V_{cr}	Critical mean velocity	(m/s)	LT^{-1}
v_y	Local velocity in y direction	(m/s)	LT^{-1}
v_{av}	average velocity of the water	(m/s)	LT^{-1}
w	Local velocity in z direction;	(m/s)	LT^{-1}
w_s	Fall velocity of a sediment particle	(m/s)	LT^{-1}
x	Horizontal scale	(m)	L
x, y, z	Cartesian coordinates	(m)	L
y	Flow depth; vertical scale; water depth	(m)	L
y_c	Critical depth	(m)	L
y_n	Normal depth	(m)	L
z	Distance	(m)	L
z	Bed elevation above datum	(m)	L
Z	Suspension number	(−)	−
∞	Infinite	(−)	−

LIST OF GREEK SYMBOLS

Symbol	Quantity	Unit	Dimension
α	Correction factor	(−)	−
α	Angle	Radian	−
β	Momentum correction factor (Boussinesq)	(−)	−
γ	Specific weight	(N/m^3)	FL^{-3}
γ_d	Form factor	(−)	−
γ_r	Ripple presence	(−)	−
δ_b	Saltation height	(m)	L
δ	Bed form height	(m)	L
δ	Boundary layer thickness	(m)	L
Δ	Increment	(−)	−
Δ	Relative density: $\Delta = (\rho_s - \rho)/\rho$	(−)	−
Δ	Bed form height	(m)	L
λ	Bed form length	(m)	L
ε	Mixing coefficient	(m^2/s)	L^2T^{-1}
η	Eddy viscosity	(kgs/m^2)	FTL^{-2}
σ_s	Geometric standard deviation	(−)	−
ρ	Density	(kg/m^3)	FT^2L^{-4}
ρ_s	Density of the sediment particles	(kg/m^3)	FT^2L^{-4}

(*Continued*)

Symbol	Quantity	Unit	Dimension
ρ_w	Density of water	(kg/m^3)	FT^{2}L^{-4}
θ	Shields parameter	(–)	–
θ	dimensionless mobility parameter	(–)	–
θ_{cr}	Dimensionless critical mobility parameter	(–)	–
κ	von Karman's constant; 0.4	(–)	–
ϕ	Dimensionless bed load sediment transport rate	(–)	–
ϕ	Functional relationship among non-dimensional parameters	(–)	–
ψ	Stratification correction	(–)	–
ν	Kinematic viscosity $\nu = \mu/\rho$	(m^2/s)	L^{2}T^{-1}
ν_t	Effective viscosity	(m^2/s)	L^{2}T^{-1}
μ	Absolute viscosity or dynamic viscosity	(Pa · s)	
π	Ratio of circular circumference and diameter; $\pi = 3.14159$	(–)	–
τ	Shear stress	(N/m^2 or Pa)	FL^{-2}
τ	Tractive force	(N/m^2 or Pa)	FL^{-2}
τ_{bx}	Bottom shear stress	(N/m^2)	FL^{-2}
τ_{cr}	Critical shear stress	(N/m^2)	FL^{-2}
τ'_{cr}	Critical shear stress for initiation of suspension	(N/m^2)	FL^{-2}
τ'	Grain shear stress	(N/m^2)	FL^{-2}
τ_{xy}	Effective shear stress	(N/m^2)	FL^{-2}
σ	Standard deviation	(–)	–
Σ	Summation symbol	(–)	–

Methods to Estimate the Total Sediment Transport Capacity in Irrigation Canals

A.1 INTRODUCTION

The design and operation of irrigation canals are based on the principle that no deposition or erosion occurs during certain periods. Good designs of an irrigation network as well as a well-operated irrigation system require an accurate prediction of the sediment transport of irrigation canals.

There is no universally accepted equation to determine the total transport capacity of sediment in irrigation canals. Many methods predict the sediment transport under a large range of flow conditions and sediment characteristics. The available prediction methods include the methods of Lane-Kalinski (1941), Einstein (1950), Colby (1964), Bishop et al. (1965), Bagnold (1966), Engelund-Hansen (1967), Chang et al. (1967), Toffaletti (1969), Ackers-White (1973), Yang (1979), Brownlie (1981), van Rijn (1984), etc. However, the predictability of all of them is still poor. Van Rijn (1984) stated that it is impossible to predict sediment transport with an accuracy of less than 100%. Therefore, it is quite difficult to make firm recommendations about which method to use in practice. Nevertheless, a comparison of sediment transport methods under the typical flow conditions and sediment characteristics of irrigation canals could become a powerful tool to reduce inevitable errors and inaccuracy. It is not possible to check all the existing methods to predict sediment transport. In these notes five of the most widely used methods to compute sediment transport will be presented. These methods are:

- Ackers-White;
- Brownlie;
- Engelund-Hansen;
- Van Rijn;
- Yang.

A.2 ACKERS AND WHITE METHOD

The method as developed by Ackers and White (1973) is based on flume experiments with sediment with a uniform or nearly uniform size distribution, with fully established flow conditions including bed forms, for water depths smaller than 0.4 m and for flows in the lower flow regime with Fr < 0.8. The method uses three dimensionless parameters, namely the grain size sediment parameter D_*, the mobility parameter F_{gr} and the transport parameter G_{gr} to describe the sediment transport.

- The grain size sediment parameter D_* reflects the influence of gravity, density and viscosity:

$$D_* = \left[\frac{(s-1)g}{\nu^2} \right]^{1/3} d_{35}$$ (A.1)

- The mobility parameter F_{gr} is the ratio of the shear force on a unit area of the bed to the immersed weight of a layer of grains. The equation for the transitional range $(1 < D_* < 60)$ is:

$$F_{gr} = \frac{u_*^n}{\sqrt{g d_{35}(s-1)}} \left[\frac{V}{\sqrt{32} \log\left(\frac{10h}{d_{35}}\right)} \right]^{1-n}$$ (A.2)

with:

$$n = 1.00 - 0.56 \log D_*$$ (A.3)

- The transport parameter G_{gr} is based on the stream power concept and the equation reads:

$$G_{gr} = c \left(\frac{F_{gr}}{A} - 1 \right)^m$$ (A.4)

with

$$A = \frac{0.23}{\sqrt{D_*}} + 0.14$$ (A.5)

$$m = \frac{9.66}{D_*} + 1.334$$ (A.6)

$$\log c = 2.86 \log D_* - (\log D_*)^2 - 3.53$$ (A.7)

The Ackers and White function for the total sediment transport reads as:

$$q_s = G_{gr} v d_{35} \left(\frac{V}{u_*} \right)^n$$ (A.8)

where:
$D_* = $ grain parameter (dimensionless)
$d_{35} = $ representative particle diameter (m)

h = water depth (m)
v = kinematic viscosity (m^2/s)
F_{gr} = mobility parameter (dimensionless)
A = value of F_{gr} at the nominal, initial movement
G_{gr} = transport parameter (dimensionless)
c = coefficient in the transport parameter G_{gr}
m = exponent in the transport parameter G_{gr}
n = exponent in the mobility parameter F_{gr}
u_* = shear velocity (m/s)
V = mean velocity (m/s)
q_s = total sediment transport per unit width (m^2/s)

A.3 BROWNLIE METHOD

The Brownlie method (1981) to compute the sediment transport is based on a dimensional analysis and calibration of a wide range of field and laboratory data, where uniform conditions prevailed. The transport (in ppm by weight) is calculated by:

$$q_s = 727.6c_f(F_g - F_{gcr})^{1.978}S^{0.6601}\left(\frac{R}{d_{50}}\right)^{-0.3301} \tag{A.9}$$

– grain Froude number:

$$F_g = \frac{V}{[(s-1)gd_{50}]^{0.5}} \tag{A.10}$$

– critical grain Froude number:

$$F_{gcr} = 4.596\tau_{*o}^{0.5293}S^{-0.1405}\sigma_s^{-0.1696} \tag{A.11}$$

– critical shear stress

$$\tau_{*o} = 0.22Y + 0.06(10)^{-7.7Y} \tag{A.12}$$

The value of Y follows from:

$$Y = (\sqrt{s-1}\,R_g)^{-0.6} \tag{A.13}$$

– grain Reynolds number:

$$R_g = \frac{(gd_{50}^3)^{0.5}}{31620v} \tag{A.14}$$

where:
c_f = coefficient for the transport rate
c_f = 1 for laboratory conditions
c_f = 1.268 for field conditions

F_g = grain Froude number (dimensionless)
F_{gcr} = critical grain Froude number (dimensionless)
τ_{*o} = critical shear stress
σ_s = geometric standard deviation
S = bottom slope (dimensionless)
d_{50} = median diameter (mm)
R_g = grain Reynolds number (dimensionless)
R = hydraulic radius (m)
ν = kinematic viscosity (m^2/s)

A.4 ENGELUND AND HANSEN METHOD

The method of Engelund and Hansen (1967) is based on energy considerations and a relationship between the transport and mobility parameters. The total sediment transport is calculated by:
– Transport parameter ϕ

$$\phi = \frac{q_s}{\sqrt{(s-1)gd_{50}^3}} \tag{A.15}$$

– Mobility parameter θ

$$\theta = \frac{u_*^2}{(s-1)gd_{50}} \tag{A.16}$$

The relationship between the parameters is expressed by:

$$\phi = \frac{0.1\theta^{2.5}C^2}{2g} \tag{A.17}$$

The total sediment transport is expressed by:

$$q_s = \frac{0.05V^5}{(s-1)^2 g^{0.5} d_{50} C^3} \tag{A.18}$$

where:
q_s = total sediment transport (m^3/s.m)
θ = mobility parameter
ϕ = transport parameter
V = mean velocity (m/s)
C = de Chézy coefficient (m$^{1/2}$/s)
u_* = shear velocity (m/s)
d_{50} = mean diameter (m)

The Engelund and Hansen function for the total sediment transport has been evaluated with laboratory data that were characterized by graded sediment ($\sigma_s < 1.6$) with median diameters d_{50} of 0.19 mm, 0.27 mm, 0.45 mm and 0.93 mm. The method is not recommend for the cases, in which the median size of the sediment is smaller than 0.15 mm and the geometric standard deviation is greater than 2.

A.5 VAN RIJN METHOD

The van Rijn method (1984a and 1984b) computes the total sediment transport. The total transport is the summation of the bed and suspended load transport ($q_s = q_b + q_{sus}$). The bed load transport q_b is given by the product of the saltation height, the particle velocity and bed load concentration. The method assumes that the gravity forces dominate the motion of the bed particles. The method of van Rijn has been evaluated with data from tests that had the following characteristics:
- mean velocity = 0.31–1.29 m/s
- flow depth = 0.1–1.0 m
- median diameter = 0.32–1.5 mm

The bed load transport follows from:

$$q_b = u_b \delta_b c_b \tag{A.19}$$

- Particle velocity u_b

$$u_b = 1.5 T^{0.6}[(s-1)g d_{50}]^{0.5} \tag{A.20}$$

- Saltation height δ_b:

$$\delta_b = 0.3 D_*^{0.7} T^{0.5} d_{50} \tag{A.21}$$

- Bed load concentration c_b:

$$c_b = 0.18 c_o \frac{T}{D_*} \tag{A.22}$$

with:

$$T = \frac{(u_*')^2 - (u_{*,cr})^2}{(u_{*,cr})^2} \tag{A.23}$$

$$u_*' = \frac{g^{0.5} V}{C'} \tag{A.24}$$

$$D_* = \left[\frac{(s-1)g}{v^2} \right]^{1/3} d_{50} \tag{A.25}$$

Replacing equations A.20 to A.25 in equation A.19 gives:

$$q_b = 0.053(s-1)^{0.5} g^{0.5} d_{50}^{1.5} D_*^{-0.3} T^{2.1} \tag{A.26}$$

where:
q_b = bed load transport (m²/s)
u_b = particle velocity (m/s)
c_b = bed load concentration
c_o = maximum volumetric concentration = 0.65
T = bed shear parameter
D_* = particle parameter
u'_* = bed shear velocity related to grains (m/s)
C' = de Chézy coefficient related to grains (m$^{1/2}$/s)
$C = 18 \log(12h/3d_{90})$
δ_b = saltation height (m)
d_{50} = median diameter (m)

The suspended load transport

The suspended sediment transport q_{sus}, is the depth integrated product of the local concentration and the flow velocity. The method of van Rijn (1984b) is based on the computation of the reference concentration from the bed load transport.

The field and laboratory data came from tests with the following characteristics:

- mean velocity   = 0.4–2.4 m/s
- flow depth   = 0.1–17 m
- median diameter = 0.1–0.4 mm

The suspended load transport follows from:

$$q_{sus} = FVhc_a \qquad (A.27)$$

- shape factor F:

$$F = \frac{(a/h)^{Z'} - (a/h)^{1.2}}{(1 - a/h)^{Z'}(1.2 - Z')} \qquad (A.28)$$

- suspension parameter Z

$$Z = \frac{w_s}{\beta \kappa u_*} \qquad (A.29)$$

$$\beta = 1 + 2\left(\frac{w_s}{u_*}\right)^2 \qquad (A.30)$$

- modified suspension parameter Z'

$$Z' = Z + \psi \qquad (A.31)$$

$$\psi = 2.5\left(\frac{w_s}{u_*}\right)^{0.8}\left(\frac{c_a}{c_o}\right)^{0.4} \qquad (A.32)$$

– reference concentration c_a

$$c_a = \frac{0.015 d_{50} T^{1.5}}{a D_*^{0.3}}$$ (A.33)

– reference level a

$$a = 0.5\Delta \quad \text{or} \quad a = k_s \quad \text{with } a_{min} = 0.01h$$ (A.34)

– representative particle size of suspended sediment d_s

$$d_s = [1 + 0.011(\sigma_s - 1)(T - 25)]d_{50}$$ (A.35)

where:
F = shape factor
V = mean velocity (m/s)
u_* = shear velocity (m/s)
c_a = reference concentration
h = water depth (m)
D_* = particle parameter
a = reference level (m)
Z = suspension number
Z' = modified suspension number
β = ratio of sediment and fluid mixing coefficient
ψ = stratification correction
κ = constant of von Karman
σ_s = geometric standard deviation
d_{50} = median diameter (mm)
d_s = representative particle size of suspended sediment (m)
w_s = fall velocity of representative particle size (m/s)
T = transport stage parameter
Δ = bed form height (m)
k_s = equivalent roughness height (m)
$u_{*,cr}$ = critical bed shear velocity (m/s)

A.6 YANG METHOD

The method proposed by Yang (1973) is based on the hypothesis that the sediment transport in a flow should be related to the rate of energy dissipation. The rate of energy dissipation is defined as the unit stream power and can be expressed by the velocity times slope ($V * S$). The theoretical basis for Yang's unit stream power is provided by the turbulence theory. By integrating the rate of turbulence energy production over the depth, the suspended sediment transport can be expressed as function of the unit stream power.

The total sediment transport can be expressed in ppm by mass as a function of the unit stream power by:

$$\log c_t = I + J \log \left(\frac{VS - V_{cr}S}{w_s} \right) \tag{A.36}$$

Yangs coefficients are:

$$I = 5.435 - 0.286 \log \left(\frac{w_s d_{50}}{v} \right) - 0.457 \log \left(\frac{u_*}{w_s} \right) \tag{A.37}$$

$$J = 1.799 - 0.409 \log \left(\frac{w_s d_{50}}{v} \right) - 0.314 \log \left(\frac{u_*}{w_s} \right) \tag{A.38}$$

– critical velocity for initiation of motion V_{cr}:

$$V_{cr} = 2.05 w_s \tag{A.39}$$

The total load transport is calculated by:

$$q_s = 0.001 c_t V h \tag{A.40}$$

where:
q_s = total load transport (kg/s.m)
c_t = total sediment transport expressed in ppm by mass
V = mean velocity (m/s)
V_{cr} = velocity for initiation of motion (m/s)
h = water depth (m)
S = bottom slope
I, J = coefficients in Yang's function for the total sediment transport
w_s = fall velocity (m/s)
d_{50} = median diameter (m)
u_* = shear velocity (m/s)
v = kinematic viscosity (m²/s)

Equations A.36 and A.40 have been verified with laboratory and field data from tests within the following ranges:
– sediment size = 0.15–1.71 mm
– mean velocity = 0.23–1.97 m/s
– water depth   = 0.01–15.2 m
– concentration = 10–585,000 ppm
– bottom slope = $0.043 * 10^{-3}$–$27.9 * 10^{-3}$

APPENDIX B

Methods to Predict the Friction Factor

The selection of methods to predict the friction factor in irrigation canals include:
- van Rijn (1984c);
- Brownlie (1983);
- White, Bettess & Paris (1979);
- Engelund (1966).

The methods to estimate the friction factor in irrigation canals will make use of the de Chézy coefficient. The Darcy-Weisbach friction factor f as well as the coefficients of Manning (n) and Strickler (k_s) can be applied when the following relationships are used:

$$C = \sqrt{\frac{8g}{f}} \quad \text{or} \quad C = \frac{R^{1/6}}{n} \tag{B.1}$$

where:
C = de Chézy coefficient ($m^{1/2}$/s)
f = Darcy-Weisbach's friction factor
n = Manning's coefficient ($m^{1/3}$/s)
k_s = Strickler coefficient (s/$m^{1/3}$)
R = hydraulic radius (m)
g = gravity acceleration (m/s^2)

B.1 VAN RIJN

The de Chézy coefficient depends on the type of flow regime. Based on the bed-roughness condition the flow regime in open canals can be divided in: smooth, rough and a transition regime. Roughness conditions on the bottom are simulated by using an equivalent height of the sand roughness k_s, which is equal to the roughness of sand that gives a resistance similar to the resistance of the bed form. The dimensionless value of u_*k_s/v is used as the classification parameter to distinguish the type of flow regime. Van Rijn (1993) described the type of flow regimes as presented in Table B.1.

Table B.1. Hydraulic regime types.

Type of regime	Classification parameter $u_* k_s / v$
Smooth	$u_* k_s / v < 5$
Transition	$5 < u_* k_s / v < 70$
Rough	$u_* k_s / v > 70$

Depending on the bed condition the flow regimes in irrigation canals can be determined as:

Plane bed: for plane beds (no motion) the equivalent height k_s is related to the largest particles of the bed material. Van Rijn (1982) described the equivalent roughness height as $k_s = 3 d_{90}$.

Assuming a uniform sediment size distribution in irrigation canals ($d_{90} \approx 1.5 d_{50}$), the equivalent roughness height of the sediment for plane beds can be represented by:

$$k_s = 4.5 d_{50} \tag{B.2}$$

where:
$d_{50} =$ median diameter of the sediment (m)
$k_s =$ equivalent height roughness (m)

The parameter $u_* k_s / v$ for a plane bed (bed without motion) follows from:
$k_s = 4.5 d_{50}$
$u_* =$ critical Shield shear velocity ($u_{*,cr}$)

Values of this parameter for various sediment diameters and for a plane bed are shown in Table B.2. The flow regimes that belong to the condition of a plane bed are the regimes smooth and transition.

Table B.2. $u_* k_s / v$ parameter for plane bed (no motion).

d_{50} (mm)	0.05	0.10	0.15	0.20	0.25	0.30	0.35	0.40	0.45	0.50
$u_{*(cr)} k_s / v$	2.8	5.1	8.2	11.5	14.9	18.5	21.3	25.8	30.6	35.7

Bed forms: for higher velocities, the occurrence of bed forms changes the bed roughness. The effective roughness or the total equivalent height follows from van Rijn (1984c):

$$k_s = k_s' + k_s'' \tag{B.3}$$

where:
$k_s =$ total equivalent height (m)
$k_s' =$ equivalent height related to the grain (m)
$k_s'' =$ equivalent height related to the bed form (m)

Values for k'_s and k''_s follow from the equations given by van Rijn (1982):

$$k'_s = 3d_{90} \text{ to } 4.5d_{50} \tag{B.4}$$

$$k''_s = 20\gamma_r\Delta_r\left(\frac{\Delta_r}{\lambda_r}\right) \qquad \text{(Ripples)} \tag{B.5}$$

$$k''_s = 1.1\gamma_d\Delta_d\left(1 - e^{-25(\Delta_d/\lambda_d)}\right) \quad \text{(Dunes)} \tag{B.6}$$

where:
$\gamma_r = $ ripple presence ($\gamma = 1$ for ripples alone)
$\gamma_d = $ form factor ($\gamma_d = 0.7$ for field conditions)
$\Delta_r = $ ripple height ($\Delta_r = 50$ to $200d_{50}$)
$\Delta_d = $ dune height
$\lambda_r = $ ripple length ($\lambda_r = 500$ to $1000d_{50}$)
$\lambda_d = $ dune length

The resistance due to the grain roughness is small compared to the one caused by the geometry of the bed form. Many attempts have been made to describe the geometry of ripples and dunes. Yalin (1985) described the geometry of ripples generated by a subcritical flow in channels with cohesionless and uniform bed material. The ripple length λ_r is in the interval:

$$600D \leq \lambda_r \leq 2000D \tag{B.7}$$

The "largest population" of the ripple lengths can be found within the range:

$$900d_{50} \leq \lambda_r \leq 1000d_{50} \tag{B.8}$$

A value that represents the ripple length well can be given by:

$$\lambda_r \cong 1000d_{50} \tag{B.9}$$

The ripple height can be described by Yalin (1985):

$$\Delta_r \text{ is 50 to } 200d_{50} \tag{B.10}$$

where:
$D = $ representative grain size ($D = d_{50}$)
$\lambda_r = $ ripple length (m)
$\Delta_r = $ ripple height (m)

For practical purposes it can be assumed that for $\Delta_r = 100d_{50}$ and $\lambda_r = 1000d_{50}$:

$$\frac{\Delta_r}{\lambda_r} \approx 0.1 \quad \text{for } 1 \leq D_* \leq 10 \quad \text{and} \quad T \leq 3 \tag{B.11}$$

Another bed form in the lower regime (Fr < 0.8) is the dunes. The shape of the dunes is similar to the shape of ripples, but their length and height are greater than those of ripples are. Relationships for dune length and dune height have been based on flume and field data and are given by van Rijn (1994):

$$\frac{\Delta_d}{h} = 0.11\left(\frac{d_{50}}{h}\right)^{0.3}(1 - e^{-0.5T})(25 - T) \tag{B.12}$$

$$\lambda_d = 7.3\,h \tag{B.13}$$

where:
T = excess bed-shear stress parameter
λ_d = dune length (m)
Δ_d = dune height (m)
d_{50} = median diameter (m)
h = water depth (m)

The greatest influence of k_s' (equivalent height related to the grain) on the total value of the equivalent roughness height k_s will occur for those bed forms that have the smallest equivalent height related to the bed form. The smallest height occurs for ripples and the influence of the height k_s' on the total equivalent roughness height k_s for the specific conditions of irrigation canals is approximately 1.5–2%.

$$\frac{k_s'}{k_s' + k_s''} = \frac{4.5d_{50}}{4.5d_{50} + 20\gamma_s\Delta_r\left(\frac{\Delta_r}{\lambda_r}\right)} \approx 2\% - \text{ripples occurrence} \tag{B.14}$$

For other bed forms the influence of the grains on the total roughness will be smaller than for ripples (smaller than 2%). For that reason and because the grain roughness is constant during changes of the bed form size, the grain roughness can be neglected. Hence, the equivalent height related to the bed form is recommended and for ripples the total equivalent roughness can be computed as:

$$k_s = k_s'' = 20 * 1 * 0.1 * 100 * d_{50} = 200d_{50} \tag{B.15}$$

Minimum values of the parameter u_*k_s/v for ripples in irrigation canals follow from:
• $k_s = 200\,d_{50}$;
• $u_* =$ critical Shield shear velocity u_{*cr}.

For those canals the values of the parameter u_*k_s/v are shown in Table B.3. Once the sediment on the bottom of an irrigation canal comes into motion the flow will be considered as hydraulically rough. Figure B.1 shows the types of flow regime in irrigation canals.

Table B.3. $u_* k_s / v$ parameter for ripples.

D_{50} (mm)	0.05	0.10	0.15	0.20	0.25	0.30	0.35	0.40	0.45	0.50
$u_{*(cr)} k_s / v$	499	899	1450	2037	2650	3286	3782	4590	5445	6344

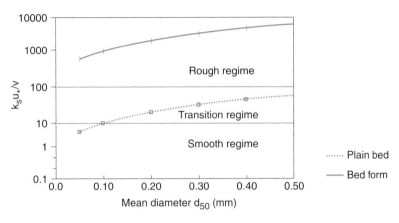

Figure B.1. Hydraulic regimes in irrigation canals.

For regimes with smooth and transitional flows, the de Chézy coefficient is a function of the flow condition only and follows from the equations given by van Rijn (1993):

$$C = 18 \log \left(\frac{12h}{3.3 \frac{v}{u_*}} \right) \qquad \text{Smooth flow regime} \qquad (B.16)$$

$$C = 18 \log \left(\frac{12h}{k_s + 3.3 \frac{v}{u_*}} \right) \qquad \text{Transition flow regime} \qquad (B.17)$$

$$C = 18 \log \left(\frac{12h}{k_s} \right) \qquad \text{Rough flow regime} \qquad (B.18)$$

where:
$C =$ de Chézy coefficient (m$^{1/2}$/s)
$h =$ water depth (m)
$v =$ kinematic viscosity (m^2/s)
$u_* =$ shear velocity (m/s)
$k_s =$ total equivalent roughness (m)

A good approximation of the de Chézy coefficient for canals with ripples is obtained by replacing the total equivalent height by the height related to the bed form k_s'' (equation B.15) and can be represented by:

$$C = 18 \log \frac{h}{200 d_{50}} \qquad (B.19)$$

The de Chézy coefficient obtained from equation B.16 to B.19 considers only the bed forms on the bottom without taking into account the friction factor of the side banks. Therefore, it is necessary to find a weighted value of the de Chézy coefficient for the friction of both bed and side banks.

B.2 BROWNLIE

Brownlie (1983) proposed a method to predict the flow depth (and therefore the friction factor) when the discharge and the slope are known. No explicit calculation of the de Chézy coefficient is proposed, but once the resistance to flow is determined (equations B.20 and B.21) then the de Chézy coefficient will be calculated by equation B.22. The Brownlie method is based on a dimensional analysis, basic principles of hydraulics and verification with a large amount of field and flume data. Step by step, the de Chézy coefficient in the lower flow regime can be predicted by using the following relationships:

$$q_* = \frac{Q}{Bg^{0.5}d_{50}^{1.5}} = \frac{q}{g^{0.5}d_{50}^{1.5}} \tag{B.20}$$

$$h = 0.372\, d_{50}q_*^{0.6539}S_o^{-0.2542}\sigma_s^{0.1050} \tag{B.21}$$

and

$$q = Ch\sqrt{hS_o} \quad \therefore C = \frac{q}{h^{1.5}S_0^{0.5}} \tag{B.22}$$

where:
Q = discharge (m³/s)
q = unit discharge (m²/s)
B = bottom width (m)
h = water depth (m)
S_o = bottom slope
d_{50} = median diameter (m)
σ_s = gradation of sediment ($\sigma_s = \frac{1}{2}(d_{84}/d_{50} + d_{50}/d_{16})$)
q_* = dimensionless unit discharge

B.3 WHITE, PARIS AND BETTESS

White et al. (1979) describe the flow resistance by the following dimensionless numbers:
– Particle size D_*

$$D_* = \left[\frac{(s-1)g}{\nu^2}\right]^{1/3}d_{35} \tag{B.23}$$

– Particle mobility F_{fg}

$$F_{fg} = \frac{u_*}{\sqrt{gd_{35}(s-1)}} \tag{B.24}$$

– Mobility parameter related to the effective shear stress

$$F_{gr} = (F_{fg} - A)\left[1.0 - 0.76\left(1 - \frac{1}{\exp(\log D_*)^{1.7}}\right)\right] + A \tag{B.25}$$

– Mean velocity:

$$V = \sqrt{32}\log\left(\frac{h}{d_{35}}\right)\left[\frac{F_{gr}\sqrt{gd_{35}(s-1)}}{u_*^n}\right]^{1/(1-n)} \tag{B.26}$$

with:

$$n = 1 - 0.56\log D_* \tag{B.27}$$

$$A = \frac{0.23}{\sqrt{D_*}} + 0.14 \tag{B.28}$$

– The de Chézy coefficient follows from:

$$C = \frac{g^{0.5}V}{u_*} \tag{B.29}$$

where:

D_* = particle size
F_{fg} = particle mobility
F_{gr} = particle mobility related to the effective shear stress
d_{35} = particle size (m)
A = initial motion parameter
n = exponent in the mobility parameter related to the effective
    shear stress
h = water depth (m)
g = gravity (m/s^2)
u_* = shear velocity (m/s)
V = mean velocity (m/s)
s = relative density
C = de Chézy coefficient (m$^{1/2}$/s)
v = kinematic viscosity (m^2/s)

B.4 ENGELUND

Engelund (1966) defined the shear stress due to skin and form resistance by:

$$\tau = \tau' + \tau'' \tag{B.30}$$

with:

$$\tau = \rho g h S \quad \text{and} \quad \tau' = \rho g h' S \quad \text{and} \quad \tau'' = \rho g h'' S \tag{B.31}$$

$$u_* = (ghS)^{0.5} \quad \text{and} \quad u_{*'} = (gh'S)^{0.5} \tag{B.32}$$

with

$$h = h' + h'' \tag{B.33}$$

Hence:

$$\left(\frac{u_{*'}}{u_*}\right)^2 = \frac{h'}{h} \tag{B.34}$$

Expressing the shear velocity in terms of the mobility parameter θ the equation becomes:

$$\frac{\theta'}{\theta} = \frac{h'}{h} \tag{B.35}$$

It was found for the lower regime that:

$$\theta' = 0.06 - 0.4\theta^2 \tag{B.36}$$

The mean velocity is calculated by:

$$\frac{V}{u_{*'}} = 6 + 2.5 \ln\left(\frac{h'}{2\,d_{50}}\right) \tag{B.37}$$

Combination of the equations will result in the de Chézy coefficient:

$$C = g^{0.5}\left(\frac{h'}{h}\right)^{0.5}\left[6 + 2.5\ln\left(\frac{h'}{2.5d_{50}}\right)\right] \tag{B.38}$$

where:
$h = h' + h'' =$ water depth (m)
$\tau = \tau' + \tau'' =$ effective shear stress (N/m^2)
$u_* =$ shear velocity (m/s)
$\theta =$ dimensionless mobility parameter
$\rho =$ density (kg/m^3)
$V =$ mean velocity (m/s)
$C =$ de Chézy coefficient (m$^{1/2}$/s)
$S =$ bottom slope

APPENDIX C

Hydraulic Design of Irrigation Canals

C.1 INTRODUCTION

In many cases, profitable and sustainable agricultural production needs an irrigation network that brings water from the source to the fields at the right place, at the right time, in the right quantity and at the right level (FAO, 1992). An irrigation network can be divided into main, secondary, sub lateral and tertiary canals. The design capacity of all the canals and appurtenant structures should be sufficient to handle the maximum envisaged flow in a convenient and reliable way. The design of the physical part of the irrigation network should enable the operation of the canal network according to the following criteria (Dahmen, 1999):

- The required flows are passed at design water levels.
- No erosion of canal bottoms and banks occurs.
- Any sediment that enters the system will be carried along the network and will discharge through outlets either to the fields, or to the natural drainage, or will settle in special designed silt traps.

The demand for irrigation water is not constant during the irrigation season, but is affected by the water requirements of the crops growing on the fields. Despite the fact that the water requirements vary during the irrigation season, the design discharge is defined as the maximum flow that can be handled in a proper way. The design discharge of a canal reach is the sum of the simultaneous, maximum flows through the outlets in this reach and the outflow into the next downstream reach plus seepage and other losses, including those due to the operation of the network. The simultaneous, maximum flow through the outlets will cover the maximum crop water requirement at field level, taking into account the number of farms irrigated at the same time and the application and distribution losses at farm and tertiary unit level. The water losses in a canal reach due to the operation are commonly included in the conveyance efficiency. When seepage losses in the canal reach are important they are separately taken into account. Water losses as a result of breakage will normally not be considered.

As stated before, the canal design should result in the right canal dimensions to convey the flows at the required water levels and with the allowable velocities. Moreover, the conveyance of the water and sediments should take place without erosion or sedimentation in the network. Based on the fact that the earthworks for an irrigation network form a considerable part of the total project costs, an optimal canal design should result in minimal earthwork, an acceptable balance between cut and fill and a minimum of borrow pits and spoil banks. If a substantial imbalance exists, the canal design has to be changed and the bottom slope, water levels and canal dimensions have to be adjusted within acceptable ranges to achieve a well-balanced earthwork.

A correct canal design should be based on the design flows of each canal reach and the final design includes:

- The layout of the canal, consisting of the horizontal and vertical alignment.
- The cross sections, including the bottom width, side slopes, water depth, freeboard and width of the embankments and the right of way.

This appendix will discuss a canal design that is based on the uniform flow equation, boundary shear stress and sediment transport relationship. The sediment transport relationship will use the principle of minimum stream power, which states that the sediment transport capacity of a canal network should be constant or non-decreasing in the downstream direction.

This approach is simple and straightforward, but there are a few limitations, namely:

- The approach does not take into account the effect of sediment size and concentration; the stream power is related to velocity and bottom slope, but basically the slope depends on the sediment size; in this method the bottom slope is adjusted to get a low boundary shear stress and the stream power can be increased without considering the sediment characteristics;
- It doe not include the effects of bed forms on the roughness; it does not account for the effect of the side slope on the effective roughness.

C.2 ALIGNMENT OF AN IRRIGATION CANAL

The layout of a canal mainly depends on the topography but is also influenced by specific geological, agricultural, engineering, and economic considerations. The alignment is largely effected by the existing topography and dominant soil conditions; areas with very steep cross-slopes or permeable soils or areas subject to landslides or that have a natural tendency for sliding should be avoided. The alignment is more economic when the water levels are kept as low as possible in irrigation canals, preferably below the ground surface. The actual water levels should never be allowed to exceed the design levels and therefore surplus water must be evacuated to drains or natural watercourses. A correct canal alignment also depends on the bottom slope and one of the main design principles is that the average slope of a main canal is more gentle than the average slope of a secondary or branch canal (Leliavsky, 1983).

The final design of the horizontal and vertical alignment should be the most economic solution in view of construction and maintenance costs and this design is not always the shortest distance between two points. The layout of an irrigation network depends on the location of the existing natural, main drains; they form the physical boundary conditions for the layout (in view of costs of land shaping). The horizontal alignment follows as much as possible the topography of the terrain and is preferably located on ridges, the main canal is situated along the watershed formed by the main ridge and the branch canals follow the less significant ridges. Curves in unlined canals should be as large as possible as they disturb the water flow and have a tendency to induce siltation on the inside and scour on the outside of the curve. The minimum radius for earthen irrigation canals depends on the size and capacity of the canal, the type of soil and the flow velocity. The recommended value for these curves is at least 8 times the design width at the water surface.

The vertical canal alignment is a compromise between the following design requirements:

- *Water levels* should be high enough to irrigate the highest areas for which irrigation is envisaged.
- *Maintenance costs* should be as low as possible; they are lower when the water level is below the ground surface. These lower water levels might hamper the illegal diversion of water by gravity.
- *Balance between cut and fill* is an economic criterion in view of the construction costs; but when this criterion results in canals in high fill it has to be mentioned that these canals are more difficult to construct and in general will lose more water by seepage.

C.3 WATER LEVELS

The water level or command is the location of the water surface in relation to a datum (reference level). The water level in a canal reach should be high enough to supply water to the highest farm plots for which irrigation is envisaged. In a gravity irrigation system water flows from the secondary canal through a turnout to a tertiary canal, and next through the farm canal system to the farm plot. The water level at the turnout, taking into account the distance to and the ground level of the critical point, should be sufficient high to convey the water to the farm plot. The design water levels of a canal follow from the *water level diagram*. This diagram presents the longitudinal profile of the ground surface and indicates all the points where water is supplied to a lower level canal or to a tertiary unit. The water level in the main canal depends on the water levels in the branch canals at the regulators or division structures and the head loss at these structures. The water level at the head of a dividing canal is lower than the water level of the parent canal due to the necessary head losses in the regulator. The final design presents the water surface preferably by a continuous straight line, located at the same levels as or above the required levels at the regulating structures. For the main as well as for the branch canals the difference between water surface and ground level should preferably not be more than 0.30 m.

C.4 EARTHWORK

The design provides the right canal dimensions to convey the discharges at the required level to meet the crop water requirements and at the same time to avoid erosion of and sedimentation on the bottom and sides of a canal. However, an important aspect in the canal design is the feasibility of the earthwork, which has a great impact on the total construction costs. An optimal canal layout will result in an proper balance between cut and fill. A most favourable canal layout has a balanced earthwork, resulting in a minimum of borrow pits and spoil banks. Moreover, a well balanced cut and fill will minimise the seepage losses, especially when the bottom level is higher than the ground surface. The bottom, but preferably the whole canal, should be in cut after stripping the top soil. Reduced seepage will reduce the danger of landslides and water logging along the canal.

When water levels have been defined to supply water to the highest points, then the bottom slopes have to be established to prevent erosion and sedimentation in the canals. A preliminary longitudinal profile, based on the required water levels and appropriate bottom slopes in view of erosion and sedimentation, will give a clear visualisation of the bottom and the water levels in each reach in order to define the earthwork. When imbalances between cut and fill exist, the level and slope may be adjusted within acceptable ranges. If the imbalance persists it will be necessary to modify the irrigation area with its tertiary blocks and to re-define the layout until a balance is reached between cut and fill. This will result in a repetitive process until an optimal layout has been established. This process can be done by the nowadays available computational techniques as many times as required.

C.5 DESIGN OF IRRIGATION CANALS

The final design of the geometry of an irrigation canal should present the bottom width B, the water depth y, the side slope m and the freeboard

F as function of the discharge Q, the bottom slope S and the roughness coefficient n. From these data follows the cross-sectional area A, the hydraulic radius R and the velocity v; the latter should be within a certain range in view of erosion and sedimentation. From the known discharge Q, the appropriate side slope m and a criterion for the optimal b/y ratio, the other canal dimensions can be derived.

The discharge Q that a cross-section with a certain area A can convey increases for a larger hydraulic radius or for a smaller wetted perimeter. A section with a certain area A and the smallest wetted perimeter will convey the largest discharge and is called the best hydraulic section. However, this does not mean that these cross-sections are always optimal. In case the water level is below the ground surface a narrow canal will give minimum excavation, for canals with the water surface above the ground level a wide canal will result in less excavation.

The flow in a canal is time and space related. The flow is steady if the water depth, discharge and cross section do not change with respect to time. When these parameters do not change with respect to space, then the flow is uniform. The most often used equations for uniform flow are the de Chézy, Strickler and Manning formulae.

These formulae are:

$$Q = A * C * R^{0.5} * S^{0.5} \quad \text{de Chézy} \tag{C.1}$$

$$Q = A * k_s * R^{2/3} * S^{0.5} \quad \text{Strickler} \tag{C.2}$$

$$Q = A * \frac{1}{n} * R^{2/3} * S^{0.5} \quad \text{Manning} \tag{C.3}$$

where:
Q = flow rate (m³/s)
A = wetted area (m²)
S = bottom slope (1)
R = hydraulic radius (m)
C = de Chézy flow resistance coefficient
k_s = Strickler smoothness factor
n = Manning roughness factor

The exponent '2/3' in the Strickler and in the Manning formula is not a constant, but it varies with the channel shape and roughness. Therefore some approximations use a modified k_s as function of the water depth y. Based on field investigations and studies the modified k_s value can be expressed as:

$$k_{s,y} = k_{s,1\,m} * y^{1/3} \quad \text{for } y < 1.0\,\text{m} \tag{C.4}$$

$$k_{s,y} = k_{s,1\,m} * y^{1/6} \quad \text{for } y > 1.0\,\text{m} \tag{C.5}$$

where:

$k_{s,y}$ = modified k_s for a canal depth y;

$k_{s,1\,m}$ = reference k_s for a canal depth of 1.0 m, based on a well maintained, large canal section.

The k_s, the smoothness factor, should be carefully estimated and is based on:

- Canal geometry, expressed by the side slope m and a b/y ratio;
- Roughness (irregularities) of the bottom and side walls; including flow obstruction due to weeds (height and density of vegetation, including flattening of it at high flows);
- Sediment transport, namely the suspended and bed load.

For any canal the roughness changes during the year and will be a minimum shortly after maintenance and a maximum when no maintenance is done. Recommended k_s values for a 1 m deep earthen canal are 12, 24, 36 and 42.5 for none, poor, fair and good maintenance conditions, respectively. The Table C.1 gives these k_s values as function of the maintenance conditions.

Table C.1. Smoothness factors as function of maintenance conditions (for earthen canals).

Maintenance	None	Poor	Fair	Good
$k_{s,1.0\,m}$	12	24	36	42.5

The recommended k_s values for a 1 m deep canal and the modified k_s value as function of the water depth (see equation C.4 and C.5) result in the graph as presented in Figure C.1.

For one value of k_s, one bottom slope S and one side slope m still an infinite range of bed width/depth ratios (b/y) can be selected. Minimizing the wetted perimeter P gives the 'best' hydraulic cross sections and the b/y ratio is:

$$\frac{b}{y} = 2 * (\sqrt{(1 + m^2)} - m) \tag{C.6}$$

where:

y = water depth (m)

b = bottom width (m)

m = cotangent of side slope (1)

The 'best' hydraulic section with the smallest perimeter P results in a maximum flow at minimum cost. However, the best hydraulic section is rarely applied, because it will not be stable due to the relatively deep excavations and a change in discharge heavily affects the water depth and the velocity. A deep section is nevertheless applied wherever possible,

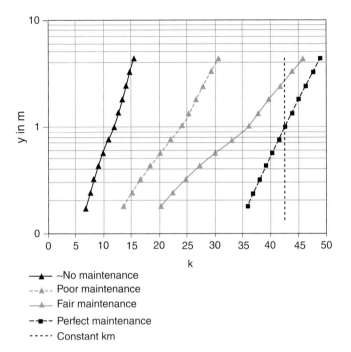

Figure C.1. Recommended k_s values for unlined irrigation canals as function of the water depth y for various maintenance conditions.

—▲— ~No maintenance
--▲-- Poor maintenance
—▲— Fair maintenance
--■-- Perfect maintenance
----- Constant km

because the expropriation costs will be less, the velocity is larger in a deep canal than in a shallow canal and the sediment transport capacity is larger in deeper canals (the transport capacity is linear with the bottom width, but exponential with the water depth). To limit the excavation and expropriation costs, the side slopes of a canal are designed as steep as possible. Soil material, canal depth and the danger of seepage determine the maximum slope of a side slope that is stable under "normal conditions", also against erosion. For deep excavations an extra berm can be included to improve the stability of the slope. Table C.2 gives some recommended values for the side slope.

Table C.2. Side slopes in canals.

Material	m
Rock	0.0
Stiff clay	0.5
Cohesive medium soils	1.0–1.5
Sand	2.0
Fine, porous clay, soft peat	3.0

The top width of canal embankments depends on the soil type, the side slope and special requirements in view of maintenance and operation.

Water levels may rise above the design water level due to deterioration of the canal embankment, a sudden closure of a gate or unwanted drainage inflows. A freeboard, being the distance between the design water level and the canal bank, is provided to safeguard the canal against overtopping these unexpected water level fluctuations and wave actions. USBR recommends a freeboard $F = C * y$, where the coefficient C varies between 0.5 to 0.6 with a minimum of the 0.15 or 0.20 m.

The bed-width/depth ratio of a small earthen irrigation canal is often close to unity; but the ratio gradually increases for larger canals. Echeverry (1915) relates this ratio to the area of the cross section, which results in:

$$\frac{b}{y_{recom}} = 1.76 * Q^{0.35} \quad \text{for } Q > 0.20 \, \text{m}^3/\text{s} \tag{C.7}$$

$$\frac{b}{y_{recom}} = 1.0 \qquad \text{for } Q \leq 0.20 \, \text{m}^3/\text{s} \tag{C.8}$$

Depending on the size of the canal, the bottom width is often rounded to a multiple of a metre or half metre. The design of a canal cross-section for a given discharge, a roughness k and a bed slope S requires iterative computations.

Re-writing the Strickler equation gives:

$$Q = A * k_s * R^{2/3} * S^{1/2}$$
$$A = (b + my)y = (b/y + m)y^2 \tag{C.9}$$
$$P = \left[\frac{b}{y} + 2\sqrt{1 + m^2}\right]y \quad \text{for } n = b/y$$

$$Q = k_s * S_o^{1/2} * \frac{(n + m)^{5/3}}{(m + 2\sqrt{(1 + m^2)})^{2/3}} \frac{y^{10/3}}{y^{2/3}} \tag{C.10}$$

$$y^{8/3} = \frac{Q}{k_s * f * S^{1/2}} \tag{C.11}$$

$$f = \frac{(n + m)^{5/3}}{(n + 2\sqrt{(1 + m^2)})^{2/3}} \tag{C.12}$$

$$\frac{b}{y_{recom}} = 1.76 * Q^{0.35} \quad \text{for } 0.2 < Q < 10 \, \text{m}^3/\text{s}$$
$$\tag{C.13}$$
$$\frac{b}{y} = 1 \qquad \text{for } Q < 0.2 \, \text{m}^3/\text{s}$$

$$k_s = k_{s,1\,m} * y^a \quad \begin{array}{l} a = 1/3 \quad \text{for } y < 1.0 \, \text{m} \\ a = 1/6 \quad \text{for } y > 1.0 \, \text{m} \end{array} \tag{C.14}$$

The design steps include:

- Assume a value for an initial Strickler k_s according to the maintenance condition and find a first approximation of y and b;
- Next re-compute the design values to obtain the proper values of b and k_s;
- Next fine-tune the design in view of a small adjustment of the design width b_{des}, i.e. a value that can be constructed (e.g. width on a half or whole meter). This fine-tuning has a minor impact on the final results, unless a completely different value from the originally suggested b is used.

When using the de Chézy equation in stead of the Strickler equation, the approach is the same as before.

The discharge for uniform flow is:

$$Q = A * C * R^{0.5} * S_o^{0.5} \quad \text{de Chézy} \tag{C.15}$$

$$C = 18 \log \frac{12R}{k + \delta/3.5} \quad \text{White-Colebrook} \tag{C.16}$$

$$\delta = 11.6 \frac{v}{v_*} \tag{C.17}$$

$$v_* = \sqrt{g * R * S} \tag{C.18}$$

where:
$C =$ de Chézy's coefficient (m^2 s^{-1})
$k =$ length characterizing the roughness $= 1/2a$ (Nikuradse)
$\delta =$ thickness of the laminar sub-layer
$v_* =$ shear velocity

Re-writing the Chézy equation gives:

$$Q = A * C * R^{0.5} * S_o^{0.5}$$
$$A = (b + my)y = (b/y + m)y^2 \tag{C.19}$$
$$P = \left[\frac{b}{y} + 2\sqrt{1 + m^2} \right] y$$

$$Q = CS_o^{0.5} \frac{(b/y + m)^{3/2} y^3}{(b/y + 2\sqrt{(1 + m^2)} y)^{1/2}}$$
$$\text{for } n = b/y: \ Q = CS_o^{0.5} \frac{(n + m)^{3/2} y^3}{(n + 2\sqrt{(1 + m^2)} y)^{1/2}} \tag{C.20}$$

$$y^{5/2} = \frac{Q(n + 2\sqrt{(1 + m^2)})^{1/2}}{CS_o^{0.5}(n + m)^{3/2}} = \frac{Q}{C * S^{0.5} * f} \tag{C.21}$$

$$f = \frac{(n+m)^{3/2}}{(n+2\sqrt{(1+m^2)})^{1/2}} \tag{C.22}$$

$$\frac{b}{y_{recom}} = 1.76 * Q^{0.35} \quad \text{for } 0.2 < Q < 10\,\text{m}^3/\text{s}$$

$$\frac{b}{y} = 1 \qquad\qquad \text{for } Q < 0.2\,\text{m}^3/\text{s} \tag{C.23}$$

The design steps include:
- Assume a value for the wall roughness (Nikuradse value) and find a first approximation of C, y and b;
- Next re-compute the design values to obtain the proper values of b and C;
- Next fine-tune the design in view of a small adjustment of the design width b_{des}, i.e. a value that can be constructed (e.g. width on a half or whole meter). This fine-tuning has a minor impact on the final results, unless a completely different value from the originally suggested b is used.

The values found in the previous design steps have to be checked in view of the erosion and sedimentation considerations, that will be discussed in the next sections.

C.6 BOUNDARY SHEAR STRESSES

The minimum velocity in a canal is that velocity that will not induce siltation and that will reduce weed growth and health risks. A velocity of 0.30 m/s is a minimum velocity in large canals; smaller velocities result in uneconomically large sections. The maximum permissible velocity should not cause erosion of the bottom and side slopes. The maximum velocity follows from the critical shear stress, which preferably should be less than 3–5 N/m² at the bottom. This shear stress is strongly influenced by the fact that:
- The resistance to erosion increases when smaller particles are washed out.
- An aged canal has more resistance to erosion than a newly constructed one.
- Colloidal matter in the water will result in cohesion of the particles that form the boundaries, resulting in an increased resistance.
- A higher ground water table than the canal water level will decrease the resistance.

According to investigations the maximum tractive force are at:
- the bottom for $b/y > 4$ to 5: $\tau = \rho * g * y * S$;
- the banks for $b/y > 2$ to 3: $\tau = 0.75 * \rho * g * y * S$.

The shear stress of the banks will not be considered unless the bed and sides are covered by coarse, non-cohesive material; in that case the angle of repose should be included.

The boundary shear stress at the bottom is:

$$\tau_{max} = c * \rho * g * y * S \ (\text{N/m}^2) \tag{C.24}$$

where:
$c =$ correction factor for various b/y ratios
$\rho =$ density of water (kg/m^3)
$g =$ acceleration due to gravity (9.81 m/s^2)
$y =$ water depth (m)
$S =$ bottom slope (m/m)

The correction factor c is a function of the b/y ratio:

$$c = 0.77 * e^{(b/y)*0.065} \quad \text{for } 1 < b/y < 4 \tag{C.25}$$

$$c = 1 \qquad\qquad\qquad \text{for } b/y \geq 4 \tag{C.26}$$

Field studies on very coarse material have shown that the "critical shear stress", above which motion of particles will start, is approximately $0.94 * D_{75}$ (N/m^2), where D is in mm. For the design a boundary shear stress of $0.80 * D_{75}$ or $0.75 * D_{75}$ is recommended.

The maximum boundary shear stress *in normal soils in normal canals and under normal conditions* can be between 3 and 5 N/m^2. The USBR recommendations for fine material are given in Table C.3. In case that the shear stress is too high the most sensitive factor to reduce the stress is a gentler bottom slope.

Table C.3. Recommended critical boundary shear stress (N/m^2).

D_{50} (mm)	Clear water	Light load	Heavy load
0.1	1.20	2.40	3.60
0.2	1.25	2.49	3.74
0.5	1.44	2.64	3.98
1.0	1.92	2.87	4.31
2.0	2.88	3.83	5.27
5.0	6.71	7.90	8.87

The design values of critical boundary shear stress are established for straight canals, and should be reduced for sinuous canals. See Table C.4 for the reduction in percentage.

Table C.4. Reduction of the boundary shear
stresses in non-straight canals.

Slightly sinuous	10%
Moderate sinuous	25%
Very sinuous	40%

C.7 SEDIMENT TRANSPORT CRITERIA

The velocity in a canal should be large enough to prevent siltation of suspended sediment. The conveyance of suspended sediment through the whole system assumes a concentration of very fine particles in suspension, which is almost evenly distributed over the vertical of the turbulent water. De Vos (1925) has stated that the relative transport capacity (T/Q) is proportional to the average energy dissipation per unit of water volume.

$$\frac{T}{Q} \propto \rho_w * g * v_{av} * S \tag{C.27}$$

where:
T/Q = relative transport capacity
ρ_w = density of water (kg/m³)
g = acceleration due to gravity (m/s²)
v_{av} = average velocity (m/s)
S = bottom slope (m/m)

From energy considerations follows that sediment particles will be transported by the water in case:

$$w \leq \frac{\rho_w * v_{av} * S}{\rho_s - \rho_w} \quad \text{or} \quad w \leq \frac{v_{av} * S}{\Delta} \tag{C.28}$$

where:
w = fall velocity (m/s)
ρ_s = density of the sediment particles (i.e. 2600 kg/m³)
ρ_w = density of the water (kg/m³)
Δ = relative density

To convey sediment in suspension the hydraulic characteristics of the canal system should remain constant or should not decrease in downstream direction. This means that $\rho_w * g * v_{av} * S$ (W/m³) should be constant or non-decreasing in downstream direction.

In a wide canal the average velocity can be expressed by $v_{av} = C * (y * S)^{0.5}$, in which C is a general smoothness factor. From this velocity follows the criterion for the continued conveyance of suspended material, namely $y * S^{1/3}$ should be constant or non-decreasing in downstream direction.

C.8 TRANSPORT OF THE BED MATERIAL

In general the bed load particles in canals have a diameter of $D > 50 * 10^{-3} - 70 * 10^{-3}$ (mm) and their transport on or above the bottom is determined by the shear velocity (v_*) with $v_* = (g * y * S)^{0.5}$.

The criterion for continued conveyance of the bed load depends on the water and sediment transport formulas, for example Einstein-Brown, Engelund Hansen, Ackers-White, Brownlie, etc.

The relative transport capacity is:

$$\frac{T}{Q} \propto \frac{b * y^3 * S^3}{b * y * y^x * S^z} \tag{C.29}$$

Where x and z are exponents depending on the choice of the water transport relationship.

The criterion for the continued movement of suspended load for a wide canal gives that $yS^{1/3}$ should be constant or non-decreasing. To prevent erosion the boundary shear stress should be constant. From this follows that $y * S$ should be constant (or non-decreasing). Therefore it would be expected that for sediment laden water that only transports bed load, the criterion for continued conveyance of the sediments should be some where in between these two criteria. The criterion to convey non-suspended (bed) material depends strongly on the water and sediment equations. The best criterion is that the relative transport capacity for bed load should be non-decreasing or in the case of possible erosion, should remain constant. The numeric approximation for this transport is that $y^{1/2} * S$ is constant or non-decreasing.

To prevent erosion the boundary shear stress should be constant, which gives that $y * S$ should be constant (or non-decreasing). Therefore it would be expected that for sediment laden water that transports only bed load, the criterion for continued conveyance of the sediments is some where in between these two values. The criterion to convey any non-suspended material depends strongly on the water and sediment equations; the best criterion would be that the relative transport capacity for bed load should be non-decreasing, or in the case of possible erosion, should remain constant. The best numeric approximation for the conveyance of non-suspended material is that $y^{1/2} * S$ is constant or non-decreasing.

C.9 FINAL REMARKS

The design of a canal section starts from a given discharge Q and looks for the following unknown canal characteristics:
1. the bottom width b;
2. the side slope m;
3. the roughness (smoothness) of the canal, k_s or C;
4. the water depth y;
5. the range for the bottom slope S_o;
    a. the minimum recommended slope;
    b. the maximum allowable slope.

The steps for the design of a canal as discussed in the previous section include:
1. the equation for uniform flow;
2. a side slope m that depends on geological criteria;
3. a value of k_s that is a function of the maintenance condition and the water depth y;
4. the recommended value for the b/y-ratio;
5. the sediment aspects:
    a. the criterion in view of the erosion (tractive force);
    b. the criterion to prevent sedimentation by controlling the relative transport capacity.

This iterative design process can be easily used for the development of a spreadsheet or a Visual Basic programme that uses all the presented criteria and equations, including the criteria for bed erosion and sediment transport. UNESCO-IHE has developed programs like Candes and Sysdes for the design of a single canal reach or a canal network with several sections, respectively.

After finalizing the design of the canal, next the model SETRIC can be applied to evaluate the design and to analyze the alternatives. The model can also be used as a decision support tool in the operation and maintenance of the irrigation network and to determine the efficiency of sediment removal facilities in the system. In addition to that the model will be helpful for the training of engineers to enhance their understanding of sediment transport in irrigation canals.

C.10 COMPUTER AIDED DESIGN OF CANALS

At present various stand-alone programs are available for the design of irrigation canals that allow the designer to evaluate a specific part of the design. In the past years UNESCO-IHE has looked for an integrated tool that combines some of the solitary programs and that results in a faster analysis of alternatives, saving time and helping to find the most feasible design. The time saved in the analysis of different canal alignments could be used for an in-depth analysis of the final selected alternative.

The newly developed tool has three modules, the first one looks for the canal alignment within the area with the least cost by using all the information produced by GIS programs (maps, tables, etc.). The second module performs the hydraulic design, including the introduction of the energy and sediment transport concepts, which avoids sedimentation and erosion in the canal network. Another feature of the tool is the possibility to export the design results of the canal design to DUFLOW (unsteady flow program) and SETRIC (sediment transport program) for further analysis of the operation and maintenance. The third module connects the results of the previous two modules to calculate the volumes of earthwork (mass diagram). As mentioned before, these volumes have a large impact on the costs and therefore it is important to optimise the design at this stage.

Tests with the new tool and comparison of the model results with the results obtained from other studies have clearly shown the applicability of GIS technology and CAD principles in engineering design. Moreover, the files with the results of the model can be easily exported to the models DUFLOW and SETRIC; this transfer of the data results in a more powerful application of the design tool. In this way the tool gives the chance to test the influence of various operation scenarios on the proposed hydraulic design. SETRIC can investigate whether the design shows the expected behaviour of the sediment in the canal network; whether all the sediment that enters the system will be transported through the whole canal system. Moreover, it will be possible to investigate and to optimise the operation and maintenance by the simulation of different predefined management scenarios.

Description of the Main Aspects of the Regime Theory

For the regime theory a set of simple, but empirical equations are available. The equations are derived from observations of alluvial canals that are relatively stable or in regime (HR Wallingford, 1992). The regime method considers the three canal characteristics, namely the perimeter of the canal, the amount of water and the amount of sediment flowing in the canal as a whole and attempts to derive the features of a stable (non silting/ non scouring) canal primarily on the basis of empirical studies of the interaction of the above-mentioned factors (Naimed, 1990). The regime theory is completely empirical and is based on measurements and observations of canals and rivers in regime. The theory originates from India and Pakistan, where most of the canal designs follow this theory, especially the Lacey regime theory. The main equations for the design of stable canals are based on field observations and experiences that are still collected and evaluated. The equations of the regime theory present a long-term average profile rather than a instantaneously variable state. Therefore, they specify the natural tendency of channels that convey sediment within alluvial boundaries to obtain a dynamic stability.

The expression canals in regime means that the canals do not change over a period of one or several typical water years. Within this period, scour and deposition will occur in the canal, but they do not interfere with canal operation. The regime theory can only provide some approximate design values as the equations are based on observations of many canals in regime. These canals have received different amounts of water and sediment from a variety of rivers that have different sediment loads and characteristics. Moreover, some of the canals in regime may have sediment excluders or ejectors at the head works that have influenced the erosion and deposition. Nevertheless, the experience obtained from the design and operation of these canals give some guidance for the design of stable channels with erodible banks and sediment transport. However, the regime theory should be applied in a very careful way, especially when the design involves a canal network with highly time-dependent operational regimes as practised in many irrigation systems at present (Bruk, 1986).

Main equations of the regime theory as given by Lacey
(Henderson, 1966)

SI-units	foot-second units	
$f = \sqrt{2520d}$	$f = 8\sqrt{d}$	(D.1)
$P = 4.836\sqrt{Q}$	$P = 2.67\sqrt{Q}$	(D.2)
$v = 0.6459\sqrt{fR}$	$V = 1.17\sqrt{fR}$	(D.3)
$S_o = 0.000315\dfrac{f^{5/3}}{Q^{1/6}}$	$S_o = \dfrac{f^{5/3}}{1750Q^{1/6}}$	(D.4)

f = silt factor for a sediment size d	f = silt factor for a sediment size d
d = sediment size (m)	d = sediment size (inch)
Q = discharge (m³/s)	Q = discharge (cusec)
P = wetted perimeter (m)	P = wetted perimeter (foot)
v = mean velocity (m/s)	v = mean velocity (foot/s)
R = hydraulic radius (m)	R = hydraulic radius (foot)
S_o = bottom slope	S_o = bottom slope

According to Lacey (1958) these equations are applicable within the following range of parameters:
- Bed material size   0.15–0.40 mm
- Discharge   0.15–150 m³/s (5–5000 cfs)
- Bed load   small
- Bed material   non-cohesive
- Bed form   ripples

The Lacey equations for the design of canals for given d and Q follow the next steps:
1. $f = (2520 * d)^{0.5}$ (equation D.1);
2. Determine $P = 4.836(Q)^{1/2}$ (equation D.2);
3. From $v = 0.6459(fR)^{1/2}$ (equation D.3), $R = A/P$ and $A = Q/v$ follows the area of the cross section $A = 1.3383 * ((P/f)^{0.5} * Q)^{2/3}$
4. Next, find the bottom slope $S = 0.000315 f^{5/3}/Q^{1/6}$ (equation D.4);
5. From the given P and A the channel section is found as a trapezoidal channel with assumed side slope (m) of $1\,V:2\,H$.
6. $y = P + (P^2 + 4A(m - 2(1 + m^2)^{0.5}))^{0.5}/(2(m - 2(1 + m^2)))$
   $B = P - 2y * (1 + m^2)^{0.5}$
   $R = A/P$

Table D.1 gives some examples of the design of earthen canals for two sediment diameters ($d = 0.0004$ m and $d = 0.00015$ m) and two side slopes ($m = 2$ and $m = 1.5$) and several discharges Q according to the Lacey method. The results are also presented in the Figure D.1 and D.2.

Table D.1. Example of the design of earthen canals for two sediment diameters and two side slopes according to the Lacey method.

SI-units

Lacey regime theory

Q	d in m	m	f	S_o	P	v	A	R	B_o	y	k_s
2.5	0.00040	2	1.00399	0.000272	7.646	0.515	4.850	0.634	3.662	0.891	42.3
3	0.00040	2	1.00399	0.000264	8.376	0.531	5.646	0.674	4.224	0.928	42.5
4	0.00040	2	1.00399	0.000252	9.672	0.557	7.176	0.742	5.223	0.995	42.9
6	0.00040	2	1.00399	0.000235	11.846	0.596	10.060	0.849	6.912	1.103	43.4
8	0.00040	2	1.00399	0.000224	13.678	0.626	12.785	0.935	8.351	1.191	43.7
10	0.00040	2	1.00399	0.000216	15.293	0.649	15.398	1.007	9.631	1.266	44.0
12	0.00040	2	1.00399	0.000210	16.752	0.669	17.925	1.070	10.797	1.332	44.2
15	0.00040	2	1.00399	0.000202	18.730	0.695	21.588	1.153	12.388	1.418	44.5
2.5	0.00015	2	0.61482	0.000120	7.646	0.438	5.712	0.747	2.006	1.261	48.5
3	0.00015	2	0.61482	0.000117	8.376	0.451	6.649	0.794	2.700	1.269	48.7
4	0.00015	2	0.61482	0.000111	9.672	0.473	8.450	0.874	3.782	1.317	49.1
6	0.00015	2	0.61482	0.000104	11.846	0.506	11.847	1.000	5.486	1.422	49.7
8	0.00015	2	0.61482	0.000099	13.678	0.531	15.056	1.101	6.897	1.516	50.1
10	0.00015	2	0.61482	0.000095	15.293	0.551	18.133	1.186	8.141	1.599	50.4
12	0.00015	2	0.61482	0.000093	16.752	0.568	21.108	1.260	9.270	1.673	50.7
15	0.00015	2	0.61482	0.000089	18.730	0.590	25.422	1.357	10.807	1.772	51.0
2.5	0.00040	1.5	1.00399	0.000272	7.646	0.515	4.850	0.634	4.693	0.819	42.3
3	0.00040	1.5	1.00399	0.000264	8.376	0.531	5.646	0.674	5.276	0.860	42.5
4	0.00040	1.5	1.00399	0.000252	9.672	0.557	7.176	0.742	6.318	0.930	42.9
6	0.00040	1.5	1.00399	0.000235	11.846	0.596	10.060	0.849	8.087	1.042	43.4
8	0.00040	1.5	1.00399	0.000224	13.678	0.626	12.785	0.935	9.597	1.132	43.7
10	0.00040	1.5	1.00399	0.000216	15.293	0.649	15.398	1.007	10.938	1.208	44.0
12	0.00040	1.5	1.00399	0.000210	16.752	0.669	17.925	1.070	12.159	1.274	44.2
15	0.00040	1.5	1.00399	0.000202	18.730	0.695	21.588	1.153	13.823	1.361	44.5
2.5	0.00015	1.5	0.61482	0.000120	7.646	0.438	5.712	0.747	3.856	1.051	48.5
3	0.00015	1.5	0.61482	0.000117	8.376	0.451	6.649	0.794	4.427	1.095	48.7
4	0.00015	1.5	0.61482	0.000111	9.672	0.473	8.450	0.874	5.441	1.173	49.1
6	0.00015	1.5	0.61482	0.000104	11.846	0.506	11.847	1.000	7.155	1.301	49.7
8	0.00015	1.5	0.61482	0.000099	13.678	0.531	15.056	1.101	8.615	1.404	50.1
10	0.00015	1.5	0.61482	0.000095	15.293	0.551	18.133	1.186	9.912	1.492	50.4
12	0.00015	1.5	0.61482	0.000093	16.752	0.568	21.108	1.260	11.093	1.570	50.7
15	0.00015	1.5	0.61482	0.000089	18.730	0.590	25.422	1.357	12.704	1.671	51.0

Example
- *Given is a canal with a discharge $Q = 10\,m^3/s$ and sediment with size $d = 0.001\,m$*
- *Hence, $f = 1.58745$ and $P = 15.293\,m$*
- *The bottom slope $S_o = 0.4636\,m/km$*
- *After trial and error follows $v = 0.76\,m/s$, $A = 13.22\,m^2$ and $R = 0.86\,m$*
- *For a side slope $m = 2$ and for the above given data follows:*
  $B_o = 10.65\,m$ and $y = 1.04\,m$.
- *Using Manning gives $n = 0.026$ and according to Strickler $k_s = 38.7$*

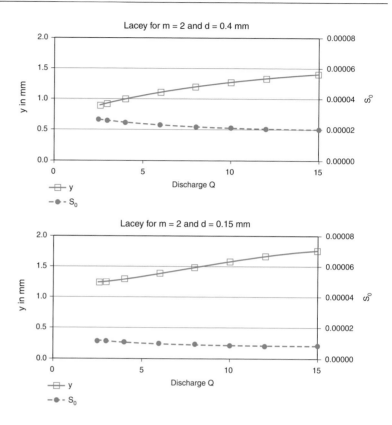

Figure D.1. Example of the design of an earthen canal according to the Lacey method for $m = 2$ and for $d = 0.04$ mm and for $d = 0.15$ mm.

D SOME REGIME CONSIDERATIONS

D.1 SEDIMENTS

In general, irrigation canals have a less steep bottom slope, smaller cross sections and discharge than the rivers, which supply the irrigation water to the canal network. This means that the canals have a smaller sediment transport capacity than the parent river. Therefore, an important aspect in the design and operation of these canals is to consider sediment exclusion at the head works or sedimentation in silting ponds. The location and crest level of canal regulators at the head works should limit the sediment entry into the canal network.

D.2 MATURING OF CANALS

The maturing of canals is an important aspect in view of the operation and maintenance of canals. The canal design according to the regime theory will be based upon the material transported by the river and not on the local soils that form the bed. The local soils might be different from the sediment transported by the river and only after a few years of operation, the bed material will be adjusted to the sediment entering the canal. The canal will then show a quasi-equilibrium condition. The maturing of the canal includes the following negative aspects:

- Development of aquatic weeds on the side-slopes;
- Adjustments of the canal to the newly developed longitudinal slope;
- Changes and improvements of off-takes, irrigation turnouts and bifurcation structures to adjust their levels to the changes in the canal after maturing.

Stable alluvial canals are characterized by regular side slopes, which are developed by the deposition of fine silt and clay particles. The side slopes are generally much less permeable than the bed and may reduce the seepage losses through the sides. They also have a higher erosion resistance and limit any tendency to widen the canal.

D.3 SLOPE ADJUSTMENTS

The bottom slope of a canal after maturing will differ from the design and the difference can be significant for long canals. Slope adjustments are feasible by the construction of drop structures at regular intervals. During the maturing process the crest levels of the drop structures have to be raised or lowered to accommodate the slopes. The drops often form a part of bifurcation structures

D.4 DIVERSION OF THE SEDIMENT

The distribution or division of the sediment over the canal branches is an important aspect in the design and operation of alluvial canals. The distribution of the flow velocity and of the sediment concentration over the flow depth is different and therefore, it is realistic to assume that the division of water and sediment at bifurcations is not proportional. Some canals will receive a larger or smaller concentration of sediment than the concentration conveyed by the parent canal. Remodelling of the bifurcation structures might be necessary to adjust the distribution of the sediment to the branches.

The aim of any irrigation network is to convey the water as well as the sediment through the farm turnouts to the fields. The sediment transport depends on the design and location of the turnout structures in relation to the canal bed. The discharge capacity of a turnout depends on the command area downstream of that turnout. When the command area increases with the development of new irrigation canals, the discharge has to be increased. Turnouts on irrigation canals are relatively inexpensive (adjustable pipe turn outs) and when the canals mature and the system attains equilibrium, the pipe turnouts can be replaced by (semi-permanent) modules. The crest level of the turnouts will then be based on sediment transport considerations as well as on the discharge variation in the canals. In smaller canals, the turnouts are important to the geometry of the canals. It has been found from intensive observations that the average bed level of distributaries adjusts to the mean crest level of these turnouts. In unstable distributaries, which may be silted up, the lowering of the turnouts is one of the recommended solutions.

D.5 MAINTENANCE ASPECTS

All irrigation canals require maintenance; earthen canal banks are constantly eroded by wind and rainfall and have to be maintained. Their stability is based upon the stability of the canal bed and the cross sections. The canal width is generally maintained by providing bank protection by permeable spurs normal to the banks. Normally, bed levels are excavated during closures if the canals silt up.

The crest of drop structures might be raised if the bed shows scouring. Maintenance of the bed elevation and canal width is expensive. Experiences show that on larger canals, bank protection is rarely practised and bed clearance is never carried out. In the smaller distributaries both the maintenance of the width and the bed elevation are often done. The berms along these canals may become a problem as they may become beyond their design width and may have to be trimmed. The canal bed level may also have to be re-excavated to near design levels. In general, the frequency of these operations, where needed, is about once in the five years. The medium size canals are more stable. In these canals like in the large canals no bank protections are practised, but the bed level may be cleared once in the ten years.

The variation in discharge in the canals is an important factor in the stability of alluvial canals. Low discharges flowing in canals with high capacities often adopt a winding thalweg that may later erode the banks by concentrating flows against the banks. For this reason the minimum discharge in mature canals is limited to 55% of the design capacity. Similarly, the raising and lowering of the canal discharge is also controlled to prevent the failure of banks due to seepage forces.

D.6 FLOW CAPACITY

In recent years, the design capacities in almost all canals have been increased to meet the growing demand for irrigation water. In some areas, this might be done by modifying the cross-sections of the existing canal and in other places by changing the structures, raising the banks and forcing the larger water supply to flow in the canal. Often this has resulted in a change of the canal regime established in the past and in unsymmetrical cross sections.

D.7 DESIGN CONSIDERATIONS

To design a straight, earthen canal that has to convey a certain amount of water and sediment three equations are needed to find the bottom width, water depth and bottom slope of the canal. However, normally there are only two equations available, namely a flow equation (Manning, Strickler or de Chézy) and a sediment transport equation. An equation to describe the shape of a stable channel is not yet available. If the tractive force along the canal perimeter is large enough to maintain sediment transport, one can rely on a kind of regime-like relationship to describe the stability of the canal (Bettes, et al., 1988).

The limitation of this theory is that it assumes that the discharge is the only factor determining the wetted perimeter of the canal while the fact is that canals with less stable banks tend to be wider than those with strong banks for the same discharge.

Simons and Albertson (1963) developed a set of regime equations based on a large data set collected from canal systems in India and North America. Five types of canals were identified from the collected data set and for each category the coefficients are given for the design of a canal as shown in Table D.2. Simons and Albertson's equations have a wide range of applicability and the data set used is related to a sediment load of less than 500 ppm (Raudkivi, 1990). The set of regime equations of Simons and Albertson (1963) is:

$$P = K_1 \sqrt{Q} \tag{D.5}$$

$$B = 0.9P = 0.92B_s - 0.61 \tag{D.6}$$

$$R = K_2 Q^{0.36} \tag{D.7}$$

$$h = 1.21R \qquad \text{for } R < 2.1 \, \text{m} \tag{D.8}$$

$$h = 0.61 + 0.93R \quad \text{for } R > 2.1 \, \text{m} \tag{D.9}$$

$$v = K_3(R^2 S)^n \tag{D.10}$$

$$\frac{C^2}{g} = \frac{v^2}{gDS} = K_4 \left(\frac{vB}{v} \right)^{0.37} \tag{D.11}$$

Table D.2. Value of coefficients in Simons and Albertson's equations for different canal types (Simons and Albertson, 1963).

Canal type	Coefficient					
	K_1	K_2	K_3	K_4	K_5	n
1. Sand bed and banks	6.34	0.4–0.6	9.33	0.33	2.6	0.33
2. Sand bed and cohesive banks	4.71	0.48	10.80	0.53	2.25	0.33
3. Cohesive bed and banks	4.0–4.7	0.41–0.56	–	0.88	2.25	–
4. Coarse non-cohesive material	3.2–3.5	0.25	4.80	–	0.94	0.29
5. Sand bed, cohesive bank and heavy sediment load*	3.08	0.37	9.70	–	–	0.29

*Sediment (mainly wash load) 2,000 to 3,000 ppm.

$$A = K_5 Q^{0.87} \qquad\qquad\qquad (D.12)$$

where:
A = cross sectional area of the flow (m^2)
B = mean canal width (m) = A/h
B_s = water surface width (m)
C = Chézy coefficient (m$^{1/2}$/s)
h = mean flow depth (m)
K_i = coefficient (i = 1 to 5) for different canal types
n = exponent
P = wetted perimeter (m)
Q = flow rate (m^3/s)
R = hydraulic mean radius (m)
S = bed slope (m/m)
V = flow velocity (m/s)
v = kinematic viscosity (m^2/s)

Regime theories are available in many areas where irrigation is practiced, but the fact that these methods are not being transformed to other places is an indication that not all the physical parameters defining the problems are correlated by the regime methods (Raudkivi, 1990).

APPENDIX E

Glossary Related to Sediment Transport

Abrasion	Frictional erosion by material transported by wind and water.
Absolute pressure	The pressure above absolute zero.
Absorbed water	Water held mechanically in a soil and possessing the same physical properties as ordinary water at the same temperature and pressure.
Accretion	Accumulation of sediment deposited by natural flow processes; increase of channel bed elevation resulting from the accumulation of sediment deposits.
Accuracy	The degree of conformity of a measured or calculated value to its definition or with respect to a standard reference.
Adequacy	The ratio of the amount delivered to the amount required.
Adjustable flow	Irrigation flow regulated at a specific discharge (flow rate).
Aggradation	Build-up or rising of channel bed due to sediment deposition.
Alignment	The course along which the centre line of a canal or drain is located.
Alluvial soil	A soil formed from deposits by rivers and streams (alluvium).
Alternate depth	In open channel flow, for a given flow rate and channel geometry, the relationship between the specific energy and flow depth indicates that, for a given specific energy, there is no real solution (i.e. no possible flow), one solution (i.e. critical flow) or two solutions for the flow depth. In the latter case, the two flow depths are called alternate depths. One corresponds to a subcritical flow, the second to a supercritical flow.
Altitude	Vertical height of a ground or water surface above a reference level (datum).
Amplitude	Half of the peak-to-trough range (or height).
Analytical model	System of mathematical equations that are the algebraic solutions of the fundamental equations.
Angle of repose	The maximum slope (measured from the horizon) at which soils and loose materials on the banks of canals, rivers or embankments stay stable.
Armouring	Progressive coarsening of the bed material resulting from the erosion of fine particles. The remaining coarse material layer forms armour, protecting further bed erosion.
Automatic control	Adjustment of the controlling element accomplished by an arrangement of equipment without attendance by an operator or continuous control of equipment without intervention of an operator to predetermined conditions.

Automatic system	System that starts and stops without human intervention in response to control, such as a switch, or a valve.
Backwater	Flow profile controlled by downstream flow conditions in a subcritical flow: e.g. a structure, change of cross-section; is commonly used for both supercritical and subcritical flow. Term backwater calculation or profile refers to the calculation of the flow profile.
Backwater calculation	Calculation of the free-surface profile in open channels.
Backwater curve	Longitudinal profile of the water surface in an open channel where the flow depth has been increased by e.g. an obstruction, increase in channel roughness, decrease in channel width or bottom slope. The water surface slope is less than the bottom slope; the curve is upstream of the obstacle that raises the water level and is concave upwards. Occasionally the term is used for other non-uniform profiles, upstream or downstream.
Bagnold	Ralph Alger Bagnold (1896–1990) – British geologist and expert on the physics of sediment transport by wind and water.
Bakhmeteff	Boris Alexandrovitch Bakhmeteff (1880–1951) – Russian expert on hydraulics, who developed the concept of specific energy and an energy diagram for open channel flows (1912).
Bank protection	the process by which the bank is protected from erosion by lining or by retarding the velocity along the bank; device to reduce scour by flowing water e.g. mattresses, groynes, revetments.
Bank, left/right-	bank to the left/right of an observer looking downstream.
Barré de Saint-Venant	Adhémar Jean Claude Barré de Saint-Venant (1797–1886), French engineer, who developed the equation of motion of a fluid particle in terms of the shear and normal forces exerted on it.
Bazin	Henri Emile Bazin: French scientist (1829–1917) and engineer, worked as an assistant of Henri P.G. Darcy.
Bed erosion	Deepening of a channel by the gradual wearing away of bed material, mainly due to the forces of flowing water.
Bed forms	Channel bed irregularity that is related to the flow conditions; features on a channel bed resulting from the movement of sediment over it. Characteristic bed forms include ripples, dunes and anti-dunes.
Bed load	Sediment transport mode in which individual particles either roll, slide or hop along the bed as a shallow, mobile layer a few particle diameters deep; mass (or volume) of course sediment (silt, sand, gravel) in almost continuous contact with the bed and transported in unit time.
Bed material	Material with particle sizes that are found in significant quantities in that part of the bed affected by transport.
Bed material load	Part of the total sediment transport, which consists of the bed material and whose rate of movement is governed by the transporting capacity of the channel.
Bed or bottom slope	Difference in bed elevation per unit horizontal distance in the flow direction.
Bed profile	Shape of the bed in a vertical plane; longitudinally or transversely, which should be stated.

Bed shear stress	The way in which currents transfer energy to the channel bed.
Berm	Horizontal edge on the side of an embankment or channel section, to intercept earth rolling down the slope, or to add strength to the construction.
Bernoulli	Daniel Bernoulli (1700–1782); Swiss mathematician, physicist and botanist who developed the Bernoulli equation: 'Hydrodynamica, de viribus et motibus fluidorum' textbook (1st publication in 1738, Strasbourg).
Bifurcation	Location where a river or channel separates in two or more reaches, sections or branches (opposite of a confluence).
Boundary conditions	Physical conditions (hydraulic and/or others) used as boundary input to physical or numerical models.
Boundary layer	The flow region next to a solid boundary where the flow field is affected by the presence of the boundary and where friction plays an essential part. A range of velocities characterizes a flow across the boundary layer region; from zero at the boundary to the free-stream velocity at the outer edge of the boundary layer.
Boussinesq	Joseph Valentin Boussinesq (1842–1929) – French expert on hydrodynamics 'Essai sur la théorie des eaux courantes' an outstanding contribution in hydraulics literature (1877).
Boussinesq coefficient	Momentum correction coefficient.
Bresse	Jacques Antoine Charles Bresse (1822–1883) – French applied mathematician and hydraulic expert, who made contributions to gradually varied flows in open channel hydraulics (Bresse 1860).
Bulk density	The mass of soil per unit of undisturbed bulk volume.
Canal	A man-made or natural channel used to convey and in some cases also to store water.
Capacity	The ability to receive or to hold power; the ability of a stream to transport water or sediment; the volume of artificial or natural reservoirs, basins.
Cartesian coordinate	One of three coordinates that locate a point in space and measure its distance from one of three intersecting coordinate planes measured parallel to one of the three straight-line axes that are the intersections of the other two planes. It is named after the mathematician René Descartes.
Caving	The collapse of a bank caused by undermining due to the wearing away of the bank by the action of flowing water.
Channel	Stream or waterway; 'channel' generally means the deep part of a river or other waterway.
Channel, open-	Longitudinal boundary consisting of the bed and banks or sides forming a passage for water with a free surface.
Channel, stable-	Channel in which the bed and sides remain stable over a significant period of time in the control reach and in which scour and deposition during floods is minimal.
Channel, unstable-	Channel in which there is frequently and significantly change in the bed and sides of a control reach.

Chézy coefficient	de Chézy coefficient is a resistance coefficient for open channel flows. Although it was thought to be a constant, the coefficient is a function of the relative roughness and Reynolds number.
Chézy, de	Antoine de Chézy (1717–1798).
Cohesion	The mutual attraction of soil particles due to strongly attractive forces. Cohesion is high in clay but significantly little in silt or sand.
Cohesionless	Referring to soil that consists primarily of silt or larger grain sizes and in which strength is directly related to confining stresses.
Cohesionless soil	Soil that has a little tendency to stick together whether wet or dry, such as sands and gravel.
Cohesive sediment	Sediment material of very small sizes (less than 50 μm) for which cohesive bonds between particles are significant and affect the material properties. The sediment contains significant proportion of clays, the electro-magnetic properties of which cause the sediment to bind together.
Cohesive soil	Very fine-grained unconsolidated earth materials for which strength depends upon moisture content.
Compound cross section	A cross section of a channel in which the width suddenly increases above a certain level.
Confidence interval	An interval around the computed value within which a given percentage of values of a repeatedly sampled variate is expected to be found.
Conjugate depth	Another name for sequent depth in open channel flow.
Continuity	The fundamental law of hydrodynamics, which states that for incompressible flows independent of time, the sum of the differential changes in flow velocities in all directions must be zero.
Control	Physical properties of a channel, which determine the relationship between stage and discharge at a location in that channel.
Control section	A section in an open channel where critical flow conditions take place resulting in a distinct relationship between water level and discharge; the concept of 'control' and 'control section' are used with the same meaning.
Coriolis	Gustave Coriolis (1792–1843) – French mathematician, who first described the Coriolis force (effect of motion on a rotating body).
Coriolis coefficient	Kinetic energy correction coefficient.
Critical depth	Depth in a canal of specified dimensions at which the mean specific energy is a minimum for a given discharge; the flow is critical.
Critical flow	Flow for which the specific energy is minimum for a given discharge; Froude number will be equal to unity and surface disturbances will not travel upstream.
Critical flow condition	The flow condition in open channel flows such that the specific energy is a minimum. Critical flow conditions occur for Fr $= 1$.
Critical slope	Slope of a channel in which uniform flow depth occurs at critical depth.
Critical tractive force	The threshold value of the tractive force of a water flow when bed material starts to move.

Cross-section	Section of a stream normal to the flow direction bounded by the wetted perimeter and the free surface.
Darcy-Weisbach friction factor	Dimensionless parameter characterizing the friction loss in a flow.
Datum	Any permanent line, plane or surface used as a reference datum to which elevations are referred.
Degradation	Lowering of channel bed elevation resulting from the erosion of sediments; downward and lateral erosion.
Density	Mass (in kg) per unit of volume of a substance.
Depth, hydraulic or mean-	Depth obtained by dividing the cross-sectional area by the free surface or top width.
Design discharge	A specific value of the flow rate, which after the frequency and the duration of exceedance have been considered, is selected for designing the dimensions of a structure or a system or a part thereof.
Design flow	Flow on which the cross sectional area of a canal or drain is determined.
Design flow depth	Depth of water in a channel when design flow moves through the cross-sectional area.
Desilting basin	Canal section with very low velocity, forming deposition of the sediment carried by the water.
Diffusion	The process whereby particles intermingle as the result of spontaneous movement and move from a region of higher concentration to one of lower concentration. Turbulent diffusion describes the spreading of particles caused by turbulent agitation.
Diffusion coefficient	Quantity of a substance that passes through each unit of cross-section per unit of time.
Dimensional analysis	Technique to reduce the complexity of a problem, by expressing the relevant parameters in terms of numerical magnitude and associated units, and grouping them into dimensionless numbers.
Discharge	Volume of water per unit time flowing along a pipe or channel, or the output rate of plant such as a pump.
Discharge regulator	Structure regulating the flow from one canal to another.
Diversion structure	Structure in a river or canal to divert all or some of the water into a (diversion) canal.
Drawdown curve	Water surface profile, when water surface slope is larger than bottom slope.
Drop	A rapid change of bed elevation, also called a step.
Dry density	Mass of soil per unit of volume on dry weight basis.
Dunes	Types of bed form indicating significant sediment transport over a sandy bed.
Dynamic equilibrium	Short-term morphological changes that do not affect the morphology over a long period.
Dynamic viscosity	The ratio between the shear stress acting along any plane between neighbouring fluid elements and the rate of deformation of the velocity gradient perpendicular to this plane.

Eddy	A vortex type of water motion, partly flowing opposite to the main current; small whirlpool; movement of flowing water in a circular direction, caused by irregularities or obstructions in canals, rivers or streams.
Eddy viscosity	Name for the momentum exchange coefficient; also called 'eddy coefficient'.
Elevation head	The vertical distance to a point above a reference level.
Energy (grade) line	Plot of the total energy head in the flow direction.
Energy gradient	Difference in total energy per unit of horizontal distance in the flow direction.
Energy head, Total-	Sum of the elevation of the free water surface above a horizontal datum in a section and the velocity head at that section.
Energy loss/head loss	Difference in total energy between two cross-sections.
Energy, Specific-	Sum of the elevation of the free water surface above the bed and the velocity head at that section.
Equity	Criterion of the share for each individual or group that is considered fair by all system members; the measure can be defined as the delivery of a fair share of water to users throughout a system; the spatial uniformity of the ratio of the delivered amount of water to the required (or scheduled) amount.
Erodible	Used for the material (soil) prone to erosion.
Erosion	The process by which soil is washed or otherwise moved by natural factors from one place to another; the gradual wearing away of canal beds by flowing water.
Erosion and scour	The detachment and transport of sediment particles because of flowing water. The processes can occur over land or within rivers and canals and ultimately result in the deposition of transported sediment at downstream locations. The movement of sediment can lead to the loss of significant soil volumes, the instability of canal beds and banks, the undercutting of bridge pier and abutment foundations, the reduction of conveyance and storage capacity, and the damaging of aquatic habitat.
Erosive	Used for the mechanism causing erosion.
Error	The difference of a measured value from its known true or correct value (or sometimes from its predicted value).
Euler	Leonhard Euler (1707–1783) – Swiss mathematician and physicist.
Event	An occurrence, which meets specified conditions, e.g. damage, a threshold rainfall, water level or discharge.
Extreme	A value expected to be exceeded once, on average, in a given (long) period of time.
Fall	A sudden drop in bed level of a stream.
Fixed-bed channel	The bed and sidewalls are non-erodible; neither erosion nor accretion occurs.
Flow	Movement of a volume of water; not to be confused with 'rate of flow' or 'discharge'.

Flow control method	In general, a regulation method for irrigation structures to maintain a specific flow condition in the irrigation system.
Flow, steady-	Flow in which the depth and velocity remain constant with respect to time.
Flow, uniform-	Flow in which the depth and velocity remain constant with respect to space.
Free flow	Flow through a canal, which is not affected by the level of the downstream water.
Free water level	Level or surface of a body of water in free contact with the atmosphere, meaning at atmospheric pressure.
Freeboard	The difference in elevation of the maximum (normal) flow line and the ground surface or bank at a canal or drain section. Also the vertical distance between the maximum water surface elevation in design and the top of retaining banks or structures, provided to prevent overtopping because of unforeseen conditions.
Free-surface	Interface between a liquid and a gas. More generally a free surface is the interface between the fluid (at rest or in motion) and the atmosphere.
Friction	Boundary shear resistance of the wetted surface of a channel, which opposes the flow of water; process by which energy is lost through shear stress.
Friction coefficient	Coefficient used to calculate the energy gradient by friction.
Froude	William Froude (1810–1879) – English naval architect and hydro-dynamicist who used the law of similarity to study the resistance of model ships.
Froude number	A dimensionless number representing the ratio of inertia forces and gravity forces acting upon water. $Fr^2 = \bar{v}^2/gy$. It is used to differentiate open channel flow regimes and to distinguish between sub-critical and super-critical flow.
Gauge	Device for measuring water level, discharge, velocity, pressure, etc., relative to a datum.
Gauge datum	Fixed plane to which the water level is related; the elevation of the zero of the gauge is related to this plane.
Gauge height	Elevation of water surface measured by a gauge.
Gradient	Measure of slope in metre of rise or fall per metre of horizontal distance; dimensionless.
Gradually varied flow	Flow characterized by relatively small changes in velocity and pressure distributions over a short distance.
Head	Potential energy of water due to its height above a datum or reference level, usually expressed in m.
Head works	The main intake structure of an irrigation system, often equipped with a weir or barrage, an intake sluice, sediment excluder or sand trap.
Head-discharge relationship	Curve or table which gives the relation between the head and the discharge in an open channel at a given cross-section for a given flow condition, e.g. steady, rising or falling.

Hydraulic	Relating to the flow or conveyance of liquids, especially water through pipes or channels.
Hydraulic jump	Abrupt, sudden change from supercritical to subcritical flow; flow transition from a rapid (supercritical flow) to a slow flow motion (subcritical flow). Giorgio Bidone published the first investigations in 1820. The present theory was developed by Bélanger (1828) and has been verified experimentally by many researchers (Bakhmeteff and Matzke 1936).
Hydraulic radius	Quotient of the wetted cross-sectional area and the wetted perimeter.
Hydrostatic pressure	Also known as the gravitational pressure, it is the pressure at a point due to the weight of the fluid above it.
Ideal fluid	Frictionless and incompressible fluid; fluid has zero viscosity: it can't sustain shear stress at any point.
International System of Units	SI = Système International d'Unités; system of units adopted in 1960 based on the metre-kilogram-second (MKS) system. It is commonly called the SI unit system. The basic seven units are: for length, mass, time, electric current, luminous intensity, amount of substance and temperature.
Irregular waves	Waves with random wave periods (in practice also heights), which are typical for natural wind-induced waves.
Karman constant	'Universal' constant of proportionality between the Prandl mixing length and the distance from the boundary. Experimental results indicate that $\kappa = 0.40$.
Karman, von	Theodore von Karman (1881–1963) – Hungarian expert on fluid-dynamics, who studied the vortex shedding behind a cylinder (Karman vortex street).
Kinematic viscosity	Dynamic viscosity divided by the fluid density.
Lagrange	Joseph-Louis Lagrange (1736–1813) – French mathematician and astronomer.
Laminar flow	Flow characterized by fluid particles moving along smooth paths in thin layers (laminas), with one layer gliding smoothly over an adjacent layer and not influenced by adjacent layers perpendicular to the flow direction. Laminar flows follow Newton's law of viscosity that relates shear stress to the rate of angular deformation: $\tau = \mu(\delta v/\delta y)$.
Level-top canal	A canal with horizontal embankments between cross-regulators to meet zero flow condition for downstream control.
Longitudinal section	Vertical section along the centre line of a canal, it shows the original and the final levels.
Main canal	Is the irrigation canal taking water from the supply (source) and conveying the water to the lateral or secondary canals.
Maintenance	Operations performed in preserving irrigation or drainage canals and drain pipes, hydraulic structures, service roads and works in good or near original conditions. Repairs are part of maintenance; the regular, continuous inspection and repair of irrigation and drainage systems.
Manning	Robert Manning (1816–1897); Chief Engineer of Public Works, Ireland. In 1889, he presented two formulae, but he preferred to use the second formula.

Mathematical model	Model that simulates a system's behaviour by a set of equations, perhaps together with logical statements, by expressing relationships between variables and parameters.
Mean depth	Average depth of a canal, being the cross sectional area divided by the surface or top width.
Mean velocity	Average velocity in a canal, being the discharge divided by the cross sectional (wetted) area.
Meandering	Single channel having a pattern of successive deviations in alignment that results in a more or less sinusoidal course.
Mixing length	The mixing length is the characteristic distance travelled by a fluid particle before its momentum is changed by the new environment; the mixing length theory is a turbulence theory developed by Prandtl (1925).
Modelling	Simulation of some physical phenomenon or system with another system believed to obey the same physical laws or rules in order to predict the behaviour of the former by experimenting with the latter.
Momentum exchange coefficient	Apparent kinematic (eddy) viscosity in turbulent flows; analogous to the kinematic viscosity in laminar flows. Momentum exchange coefficient is proportional to the shear stress divided by the strain rate. Also called eddy viscosity or eddy coefficient (Boussinesq, 1877).
Navier	Louis Marie Henri Navier (1785–1835) – French engineer who extended Euler's equations of motion (Navier 1823).
Navier-Stokes equation	Momentum equation applied to a small control volume of incompressible fluid; usually written in vector notation. Navier first derived the equation by a different method. De Saint-Venant in 1843 and Stokes in 1845 derived it in a more modern manner.
Negative surge	A negative surge results from a sudden change in flow that decreases the flow depth. It is a retreating wave front moving upstream or downstream.
Newton	Sir Isaac Newton (1642–1727) – English mathematician and physicist. His contributions in optics, mechanics and mathematics were fundamental.
Nikuradse	J. Nikuradse – German engineer who investigated the flow in smooth and rough pipes (1932).
Non uniform flow	Flow that varies in depth, cross-sectional area, velocity and hydraulic slope from section to section.
Normal depth	Uniform equilibrium open channel flow depth; depth at a given point in a canal corresponding to uniform flow, water surface and bed are parallel.
Numerical modelling	Refers to the analysis of physical processes (e.g. hydraulic) using computational models.
Off-take	Structure with or without gates, that conveys water to a secondary canal or tertiary unit.
One-dimensional flow	Neglects the variations and changes in velocity and pressure transverse to the main flow direction.
One-dimensional model	Model defined with one spatial coordinate, the variables being averaged in the other two directions.

Open canal	Natural or man-made structure that contains, restricts and directs the flow of water. The surface of the water is open to the atmosphere, and therefore, the flow is referred to as free flow. The design of canals includes the solution of relationships between bed and bank roughness, channel geometry, and flow velocity. Free surface flows are driven by gravity and they can vary in both time and space.
Open channel	Channel in which the water surface is in free contact with the air (free surface).
Overflow structure	Structure with water flowing over its crest.
Particle size distribution	The fractions of clay, silt and sand particles in a soil.
Pascal	Blaise Pascal (1623–1662) – French mathematician, physicist and philosopher, who developed the modern theory of probability. Between 1646 and 1648, he formulated the concept of pressure and showed that the pressure in a fluid is transmitted through the fluid in all directions. The unit of pressure is named after Pascal: one Pascal equals a Newton per square-metre.
Physical modelling	Investigations of hydraulic processes using a (physical) scale model.
Porosity	Percentage of the total volume of a soil occupied by air and/or water.
Positive surge	Positive surge results from a sudden change in flow that increases the depth. It is an abrupt wave front. The unsteady flow conditions may be solved as a quasi-steady flow situation.
ppm	Abbreviation for parts per million.
Prandtl	Ludwig Prandtl (1875–1953) – German physicist who introduced the concept of the boundary layer and developed the turbulent, mixing length' theory.
Precision	The degree of mutual agreement among a series of individual measurements. Precision is often, but not necessarily, expressed by the standard deviation of the measurements.
Primary canal	Canal that conveys water from the head works to the secondary canals and tertiary units.
Probability	The chance that a prescribed event will occur, represented as a pure number (p) in the range $0 < p < 1$. It can be estimated empirically from the relative frequency (i.e. The number of times the particular event occurs, divided by the total count of all events in the class considered).
Prototype	Actual structure or condition being simulated in a numerical or physical model.
Rapidly varied flow	A flow characterized by large changes over a short distance (e.g. weirs, gates, hydraulic jump).
Rating curve	Graphic or tabular presentation of the discharge or flow through a structure or channel section as a function of water stage (depth of flow).
Rayleigh	John William Strutt, Baron Rayleigh (1842–1919) – English scientist in acoustics and optics. His works are the basis of wave propagation theory in fluids.

Regression analysis	Statistical technique applied to paired data to determine the degree or intensity of mutual association of a dependent variable with one or more independent variables.
Regular waves	Waves with a single height, period and direction.
Regulator	Structure to set (regulate) water levels and/or discharges in an irrigation network.
Relative density	Density of soil material with reference to its maximum possible density for a given compaction effort. Can be expressed as a percentage of the maximum possible density, or using descriptive terms such as 'loose', 'medium' or 'dense'.
Response time	Time-lag, i.e. The time needed for a canal network to reach a new steady state after a change in water level or discharge.
Reynolds	Osborne Reynolds (1842–1912) – British physicist and mathematician who expressed the Reynolds number and stress (i.e. turbulent shear stress).
Reynolds number	Dimensionless number representing the ratio of the inertia force over the viscous force: $\mathrm{Re} = \bar{v}D/v$.
Roughness coefficient	Factor in formulae for computing the average flow velocity in open channels that represents the effect of roughness and other geometric characteristics of the channel upon the energy losses; e.g. the de Chézy, Manning or Strickler coefficients.
Roughness factor	See roughness coefficient.
Saltation	In sediment transport, particle motion by jumping and bouncing along the bed.
Sand	Sediment particles, mainly quartz, with a diameter of between 0.062 mm and 2 mm, generally classified as fine, medium, coarse or very coarse.
Sand trap	Enlargement in a channel where the velocity drops so that any sand that it carries can settle and be removed.
Scalar	A quantity that has a magnitude described by a real number and no direction. A scalar means a real number rather than a vector.
Scour	Removal of bed material by the eroding power of a flow of water; erosive action, particularly, pronounced local erosion by fast flowing water that excavates and carries away material from the bed and banks.
Sediment	Any material carried in suspension by the flow or as bed load that would settle to the bottom in the absence of fluid motion; particles transported by, suspended in or deposited by a flow.
Sediment concentration	The ratio of the mass (or volume) of the dry sediment in a water/sediment mixture to the total mass (or volume) of the suspension.
Sediment load	The amount of sediment carried/transported by running water.
Sediment transport	Movement of sediment transported in any way by a flow; from the point of transport it is the sum of suspended and bed load transported; from the point of origin it is the sum of bed material load and the wash load.
Sediment transport capacity	Ability of a stream to carry a certain volume of sediment per unit time for given flow conditions. Also called the sediment transport potential.

Sediment yield	Total sediment outflow including bed-load and suspension.
Sedimentation	Deposition of sediments in channels due to a decrease in velocity and corresponding reduction in the size and amount of sediment that can be carried.
Sedimentation basin	Basin placed at selected points along a canal to collect sand and silt.
Sequent depth	In open channel flow, the solution of the momentum equation at a transition between supercritical and subcritical flow gives two flow depths (upstream and downstream flow depths). They are called sequent depths.
Set point	The target value or desired output.
Settling basin	Small basin placed at selected points along a conduit, open canal or sub-surface drain to collect sand and silt, and also to afford an opportunity for inspection of the operation. See silt basin.
Shear strength	Ability of a material to resist forces tending to cause movement along an interior smooth surface.
Side slope	Slope of the side of a channel with the horizontal; tangent of the angle with the horizontal; the ratio of the horizontal and vertical components of the slope.
Silt	Sediment particles with a grain size between 0.004 mm and 0.062 mm, i.e. coarser than clay particles but finer than sand.
Silt basin	Small basin placed at selected points along a channel to collect sand and silt.
Silt clearance	Removal of silt deposited in a channel section above the design bed levels; the general term also includes bank trimming in the case of constriction of width by silting.
Silting	Process of accretion or rising of the channel bed by depositing of sediment in the flow. Also called 'accretion of silt'. Building of silt layers on channel sides is referred to as silting, but not as accretion.
Similitude	The correspondence between the behaviour of a model and that of its prototype, with or without geometric similarity. Scale effects usually limit the correspondence.
Simulation	Representation of a physical system by a computer or a model that imitates the behaviour of the system; a simplified version of a situation in the real world.
Slope	Inclination of the canal side from the horizontal; e.g. 1 vertical to 2 horizontal; inclination of the channel bottom from the horizontal;
Soil specific gravity	Ratio of the weight of water-free soil to its volume; bulk density; N/m^3.
Soil structure	Clustering of soil particles into units called peds or aggregates; arrangement of soil particles into aggregates that occur in a variety of recognized shapes, sizes and strengths.
Soil texture	Characterization of soils in respect of particle size and distribution; the relative proportion of the three soil fractions (sand, silt, and clay).
Specific discharge	The flow rate per unit of cross-sectional area.

Specific energy	Quantity proportional to the energy per unit mass, measured with the channel bottom as datum and expressed in meter of water.
Specific volume	The volume of a unit mass of dry soil in an undisturbed condition, equalling the reciprocal of the dry bulk density of the soil.
Steady flow	Flow that occurs when conditions at any point of the fluid do not change with the time.
Steady state	Fluid motion in which the velocity at every point of the field is independent of time in either magnitude or direction; flow condition in which the input energy equals the output energy.
Stochastic	Having random variation in statistics.
Stream function	Vector function of space and time which is related to the velocity field. The stream function exists for steady and unsteady flow of incompressible fluid as it satisfies the continuity equation. Lagrange introduced the stream function.
Streamline	A line drawn so that the velocity vector is always tangential to it (no flow across a streamline); in steady flow the line coincides with the trajectory of the fluid particles.
Subcritical flow	The flow in an open channel is subcritical if the flow depth is larger than the critical depth. A flow for which the Froude number is less than unity; surface disturbances can travel upstream.
Super-critical flow	The flow in an open channel is super-critical if the flow depth is smaller than the critical depth. A flow for which the Froude number is larger than one; surface disturbances will not travel upstream.
Surge	A surge in an open channel is a sudden change of flow depth (i.e. Abrupt increase or decrease in depth). An abrupt increase in flow depth is called a positive surge while a sudden decrease in depth is termed a negative surge. A positive surge is also called (improperly) a 'moving hydraulic jump' or a 'hydraulic bore'.
Suspended concentration	The in time-average ratio of the mass (or volume) of the dry sediment in a water/sediment mixture to the total mass (or volume) of the mixture; also average of the mean (time average) suspended concentration over the entire area.
Suspended load	Transported sediment material maintained into suspension by turbulence of the flow for considerable periods and without contact with the bed; the velocity of the load is almost the same as that of the flow; it is part of the total sediment transport.
Suspension	A two-phase system, which may consists of solid particles suspended in water.
Tertiary off-take	A discharge regulator on the secondary or primary canal to supply a tertiary unit.
Textural class	The name of a soil group with a particular range of sand, silt and clay percentages of which the sum is 100% (e.g. sandy clay: 45–65% sand, 0–20% silt, 35–55% clay).

Threshold of motion	Value at which the forces imposed on a sediment particle overcome its inertia and it starts to move.
Time lag	Time needed for a canal network to reach a new steady state after a change in water level or discharge.
Top width	Width of the channel measured across it at the water surface.
Total load (origin)	Total load comprises the 'bed material load' (including suspended load) and the 'wash load'.
Total load (transport)	The total load consists of the 'bed load' and 'suspended load' (including wash load).
Tractive force	The force exerted by flowing water on the bed and banks of a channel and tangential to the flow direction.
Tractive stress	The force per unit of wetted area that acts on the bed and bank material, trying to dislodge the material against cohesion, internal repose and gravity.
Turbulence	Flow motion characterized by its unpredictable behaviour, strong mixing properties and a broad spectrum of length scales.
Turbulent flow	Fluid particles move in very irregular paths, causing exchange of momentum from one part of the flow to another. The flow has great mixing potential and a wide range of eddy ength scales. The flow is agitated by cross-currents and eddies; any particle may move in any direction with respect to any other particle and the energy loss is almost proportional to the square of the velocity.
Two-dimensional flow	All particles are assumed to flow in parallel planes along identical paths in each of these planes. There are no changes in flow normal to the planes. An example is the flow in a wide, rectangular channel.
Uniform equilibrium flow	Flow for which the velocity is the same at every point, in magnitude and direction, at any given instant; steady uniform and unsteady uniform flow.
Uniform flow	Flow with no change in depth or any other flow characteristic (wetted area, velocity or hydraulic gradient) along a canal.
Unsteady flow	Flow in which the velocity changes, with time, in magnitude or direction.
Validation	The comparison between model results and prototype data, to validate a physical or numerical model. Validation is carried out with prototype data that are different from those used for calibration and verification of the model.
Velocity	The rate of movement at a certain point in a specified direction.
Velocity head	The energy per unit weight of water in view of its flow velocity; square of the mean velocity divided by twice the acceleration due to gravity.
Velocity, mean-	The discharge at a given cross-section divided by the wetted area at that section.
Velocity, surface-	Velocity at the water surface at a given point of a channel.
Viscosity	Property that characterizes the resistance of a fluid to shear: i.e. resistance to a change in shape or movement of the surroundings.
Void ratio	Ratio of the volume of pores to the volume of solids in a soil.

Voids	Pores or small cavities between the particles in a rock or soil mass, may be occupied by air, water or both; the spaces between stones or gravel in river structures.
Volume Control	Flow control method with the set point in the middle of the downstream or upstream canal section.
Wash load	Portion of the suspended load with particle sizes smaller than those found in the bed; in near-permanent suspension and transported without deposition; the amount of wash load transported through a section is independent of the transport capacity of the flow.
Washout or washing-out	Erosion of earth from canal banks or below structures, caused by excessive flow velocity and turbulence, accumulation of debris, burrowing animals, or bad drainage, leading to a complete breakdown of banks, structures, or lining.
Water-level	Elevation of the free water surface relative to a datum; also called stage or gauge height.
Water-level regulator	Structure to regulate the water level.
Water-surface slope	Difference in elevation of the water surface per unit horizontal distance in the flow direction.
Wave amplitudes	The magnitude of the greatest departure from equilibrium of the wave disturbance.
Wave celerity	The speed of wave propagation.
Wave frequency	The inverse of wave period.
Wave height	The vertical distance between the trough and the following crest.
Wave period	Time taken for two successive wave crests to pass the same point.
Wavelength	The straight-line distance between two successive wave crests.
Weber	Moritz Weber (1871–1951) – German Professor.
Weber number	Dimension less number characterizing the ratio of inertial forces over surface tension forces. It is relevant in problems with gas-liquid or liquid-liquid interfaces.
Wetted perimeter	The length of wetted contact between water and the solid boundaries of a cross-section of an open channel; usually measured in a plane normal to the flow direction.
Wetted surface	The surface area in contact with the flowing water in an open channel.
Width	Dimension of a channel measured normal to the flow direction.

APPENDIX F

Flow Diagrams

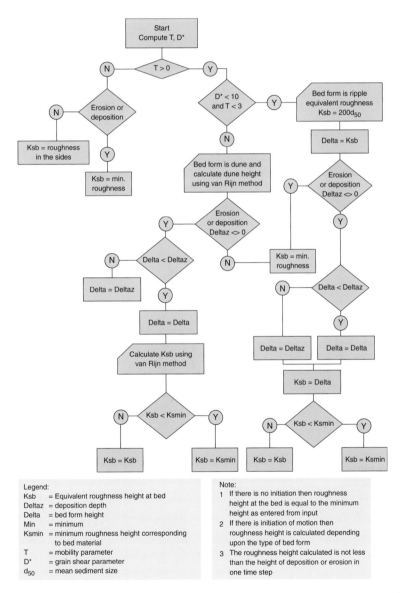

Figure F.1. Roughness prediction procedure depending upon the flow condition and sediment parameter (Paudel, 2010).

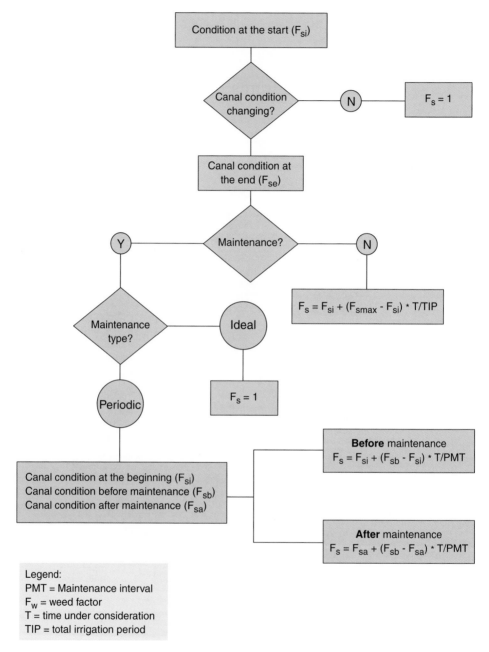

Figure F.2.   Roughness adjustment procedure for the changing canal condition with or without maintenance scenarios (Paudel, 2010).

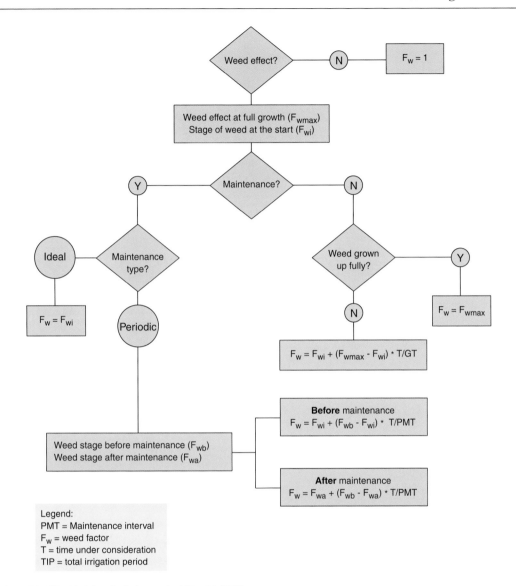

Figure F.3.   Weed height calculation method (Paudel, 2010).

Subject index